Boyer

Nutrient Cycling and Limitation

D0220872

PRINCETON ENVIRONMENTAL INSTITUTE SERIES

Series Editors
Mario J. Molina, Francois M. M. Morel,
James J. Morgan, and David Tilman

Nutrient Cycling and Limitation: Hawai'i as a Model System
by Peter Vitousek

Nutrient Cycling and Limitation

Hawai'i as a model system

Peter Vitousek

PRINCETON UNIVERSITY PRESS

OXFORD AND PRINCETON

Copyright © 2004 by Princeton University Press
Published by Princeton University Press,
41 William Street, Princeton, New Jersey 08540
In the United Kingdom: Princeton University Press,
3 Market Place, Woodstock, Oxfordshire OX20 1SY
All Rights Reserved

Library of Congress Cataloging-in-Publication Data

Vitousek, Peter Morrison.
Nutrient cycling and limitation : Hawai'i as a model system / Peter Vitousek.
p.   cm.—(Princeton Environmental Institute series)
Includes bibliographical references and index.
ISBN 0-691-11579-6 (cl. : alk. paper)—ISBN 0-691-11580-X (pb. : alk. paper)
1. Ecology—Hawaii.   2. Nutrient cycles—Hawaii.   I. Title.   II. Series.

QH198.H3V58 2004
577.314'09969—dc22   2003055551

British Library Cataloging-in-Publication Data is available

This book has been composed in Sabon

Printed on acid-free paper. ∞

pup.princeton.edu

Printed in the United States of America

1   3   5   7   9   10   8   6   4   2

To Pamela, Mat, and Liana

CONTENTS

TABLES

FIGURES

PREFACE

THIS BOOK IS a personal view of nutrient cycling and limitation in terrestrial ecosystems. My understanding of the subject has been shaped by my twenty years of research in the Hawaiian Islands, and by many friends and colleagues from a wide range of disciplines who have worked in Hawai'i. However, I do not intend this book as a synthesis of ecosystem research in the Hawaiian Islands; rather, I draw upon studies in Hawai'i to explore fundamental mechanisms involved in nutrient cycling and limitation in terrestrial ecosystems. In putting this work together, I have not hesitated to offer interpretations and applications that differ from those of the original investigators—including my own earlier interpretations. Many of my current interpretations are arguable—particularly in the final chapter—and I trust that readers will argue with them, and will offer alternative syntheses.

In presenting and discussing the results of numerous studies of Hawaiian ecosystems, I have focused on the stories that the data have to tell—most often using figures rather than tables, and avoiding the presentation of detailed statistical analyses of results. The primary data and statistical analyses generally are available in the original publications that are cited here. In addition, references, notes, and all of the numbers that go into each figure can be accessed at www.stanford.edu/vitousek/princetonbook.html.

Even more so than is usual, this book depends upon the contributions of many people. It is difficult to single out any of these friends and colleagues for particular thanks without seeming to slight essential contributions by others. Nevertheless, there are five people who must be acknowledged. Pamela Matson (Stanford University) has been my partner in carrying out this research and in developing its conceptual underpinnings; she has also managed to create space in a very busy life for me to pursue field research, analysis, and writing. Oliver Chadwick (Santa Barbara) has created and synthesized the soil science and geochemical components of the overall project, contributing his own research and drawing in other outstanding scientists from those communities. He has also generously provided unpublished data and interpretations for this book. Bill Robertson (A. W. Mellon Foundation) provided much of the organizational and moral as well as financial support that has made ecosystem research in Hawai'i much more than the sum of its parts. Doug Turner and Heraldo Farrington (Stanford) have managed the samples, laboratory analyses, data, field research, and logistics of Hawaiian research for more than a decade.

Other essential contributors to the particular research that is synthesized here included Claudia Benitez-Nelson, Jackie Heath Carrillo, Tim Crews, Lou Derry, Jim Fownes, Bob Gavenda, Robin Harrington, Lars Hedin, Darrell Herbert, Sarah Hobbie, Sara Hotchkiss, Barry Huebert, Kanehiro Kitayama, Andy Kurtz, Dieter Mueller-Dombois, Jason Neff, Becky Ostertag, Jim Raich, Ralph Riley, David Rothstein, Ted Schuur, Margaret Torn, and Kathleen Treseder. Equally fundamental contributions to research on Hawaiian ecosystems, and so to my understanding of ecosystems in general, have come from Steve Allison, Greg Aplet, Amy Austin, Teri Balser, Tracy Benning, Jon Chorover, David Clague, Susan Cordell, Carla D'Antonio, Jack Ewel, David Foote, Guillermo Goldstein, Mike Gomes, Sharon Hall, Flint Hughes, Lianne Kurina, Jack Lockwood, Kitty Lohse, Virginia Matzek, Lydia Olander, Bill Parton, Holly Pearson, David Penn, Stephen Porder, Carla Restrepo, Ann Russell, Paul Scowcroft, Lani Stemmermann, Don Swanson, Alan Townsend, and Lars Walker. Many other people have contributed to the research in other ways.

In addition, John Aber, Terry Chapin, Lars Hedin, Sarah Hobbie, Rob Jackson, Dave Karl, Becky Ostertag, Holly Pearson, Steve Perakis, and Jim Raich read earlier drafts of part or even all of this book, and I am grateful for their corrections and suggestions. Larry Bond and Doug Turner made preparation of the manuscript possible. Over the years, the research summarized here has been supported by NSF, the A. W. Mellon Foundation, the USDA NRI program, the MacArthur Foundation, the Pew Charitable Trust, NASA, and The Nature Conservancy. Access to field sites and facilities has been provided by Hawai'i Volcanoes National Park; the State of Hawai'i Divisions of Forestry and Wildlife, and of State Parks; Kahua, Parker, and Ponoholo Ranches; The Nature Conservancy Hawai'i; and the Joseph Souza Center. I thank Stanford University for providing a wonderful place to work, a great group of colleagues and students, and the flexibility to do much of my work in Hawai'i; my parents for bringing me up with the opportunity to appreciate the land; and everyone who is working to protect Hawaiian ecosystems—and thereby giving future generations the same opportunities I had.

A final word on the Hawaiian language, which contributes most of the place names referenced here. Hawaiian evolved as a spoken language— and so pronunciation is more critical to meaning than is the case in written languages. Each vowel is pronounced in Hawaiian—and in the written form, two symbols are used as guides to pronunciation. The 'okina (') indicates a glottal stop (like that between the syllables of the expression "uh-oh"), and the kahakō (as in the preceding ō) indicates a long vowel.

Nutrient Cycling and Limitation

Chapter One

INTRODUCTION

THIS BOOK brings together my two strongest interests in research: understanding nutrient cycling and limitation in terrestrial ecosystems, and understanding the ecosystems of the Hawaiian Islands. I have been fascinated by nutrient cycling and limitation since I chose to pursue a research career in ecology in 1970. I followed that fascination through analyses of temperate forest ecosystems and human influences on them, learning the dynamics of particular forests in New Hampshire, Indiana, North Carolina, and California as I moved through graduate school and a succession of university positions in those states. I worked in continental tropical forests in Costa Rica, Brazil, and Mexico, and led cross-site comparative analyses and syntheses of nutrient cycling in forest ecosystems. Shortly after moving to Stanford University in 1984, I began research in the Hawaiian Islands—where I had always intended to work, someday—and since then my research has been increasingly, now almost exclusively, focused there.

Why Hawai'i, and why nutrient cycling and limitation? The personal reasons for Hawai'i are easy to explain. I was born and brought up in Hawai'i, and most of my family remains there. It is the place I am most at home, with both land and culture. However, that sense of place is not a sufficient reason for devoting most of my research to Hawai'i, even though "a feeling for the land" does contribute substantially to that research. Still less is it sufficient reason for any agency or foundation to support my research. Part of the reason I have worked so actively in Hawai'i, and the main reason I have been able to do so, is that the Hawaiian Islands represent an extraordinary model system for the analysis of many ecosystem properties and processes. Chapters 2 and 3 develop this part of the answer to "why Hawai'i?"; they explain the concept of model systems and its application in other fields, and they show how features of islands in general—and the Hawaiian Islands in particular—make them useful model systems for answering a broad range of ecological questions.

Why I am interested in nutrient cycling and limitation is more difficult to explain—why do any of us choose the broad research areas we do? Having made my choice, though, it is easy to explain why understanding nutrient cycling and limitation is both interesting and important. The more compelling reasons include:

1. The availability and/or supply of essential nutrients demonstrably shapes the productivity, composition, diversity, dynamics, and interactions of plant, animal, and microbial populations in many terrestrial, aquatic, and marine ecosystems. N and/or P in particular are often in short supply, relative to the needs of many organisms (Schindler 1977, Vitousek and Howarth 1991, White 1993).
2. Nutrient limitation is economically important—as illustrated most directly by the fact that humanity spends tens of billions of dollars annually on fertilizer.
3. There is a fascinating question at the heart of N limitation. Biological N fixers that can draw upon the vast pool of atmospheric N_2 are ubiquitous. It seems that such fixers should have a competitive advantage in N-limited ecosystems, and as a byproduct of their activity they should boost N inputs substantially. Just that happens in most temperate freshwater ecosystems; cyanobacterial N fixation brings N supply more or less into equilibrium with the next most limiting resource (Schindler 1977). How can N limitation persist in many terrestrial ecosystems despite the presence of biological N fixers (Vitousek and Howarth 1991, Vitousek and Field 1999)?
4. The global cycles of N and P have been altered substantially by human activity. Anthropogenic N fixation for fertilizer, during fossil fuel combustion, and in legume crops is greater than natural biological N fixation on land (Smil 1990, Galloway et al. 1995, Vitousek et al. 1997a, Galloway and Cowling 2002), and the mining and mobilization of P for use in fertilizer and industrial processes exceeds the weathering of P from rocks (Carpenter et al. 1998, Smil 2000, Bennett et al. 2001). Many organisms and ecosystems now receive "unnaturally" large quantities of these elements—and function differently as a consequence. At the same time, the responsiveness of many terrestrial plants and ecosystems to the ongoing anthropogenic increase in CO_2 is constrained in part by N and/or P limitation (Schimel et al. 1997, Lloyd et al. 2001, Oren et al. 2001)—and understanding these limitations is crucial to predicting the future composition and dynamics of terrestrial ecosystems.

My goal in this book is to contribute to the understanding of nutrient cycling and limitation, using the Hawaiian Islands as a model system. Accordingly, this book is not intended as a synthesis volume for ecosystem research in Hawai'i, but rather as an analysis that makes use of unique characteristics of the Hawaiian Islands to understand the mechanisms driving nutrient availability, cycling, and limitation in terrestrial ecosystems more generally. I do not attempt to present a consensus view of ecosystem ecology, or even the consensus of my colleagues and students who have worked on Hawaiian ecosystems; this is a personal view of how I think nutrient cycling in terrestrial ecosystems works, and why. Some of the fundamental questions that I believe can be answered more straightforwardly in Hawai'i than anywhere else include:

How do biological and geochemical processes that operate on very different timescales interact to cause, sustain, or offset nutrient limitation? "Nutrient cycling" encompasses a wide range of processes that occur on very different timescales, from the turnover of available nutrients in soil solution (minutes to days, for an essential element in short supply) to the development of acidic, deeply leached tropical soils (millions of years). I will focus on four timescales here: (1) nutrient supply into biologically available pools versus organisms' demand for those nutrients, over minutes to days; (2) the cycling time of nutrients from soils to plants and back—generally years in forests—and the development of plant-soil-microbial feedbacks that can slow or speed nutrient cycling; (3) the accumulation of nutrient pools within plant biomass and soil organic matter, which can continue for decades to centuries or more, and how these pools function as sources and sinks for biologically available nutrients; and (4) the balance between inputs of elements to ecosystems and losses of elements from them, as these change over millennia to millions of years during long-term soil and ecosystem development. Despite their very different timescales, these processes interact strongly to control nutrient availability, cycling, and limitation in terrestrial ecosystems—and I believe we can understand most of these timescales and their interactions better in the Hawaiian Islands than elsewhere.

How are element inputs to and losses from terrestrial ecosystems regulated, and what are the implications for nutrient cycling and limitation? A decade ago, the prevailing conceptual model for the control of element inputs to and outputs from terrestrial ecosystems held that most inputs are controlled by processes occurring outside of ecosystems, while outputs are controlled in part by biological processes that cause nutrients in short supply to be retained within terrestrial ecosystems. Hedin et al. (1995) then suggested that nutrient limitation could be caused or sustained by losses of nutrients in forms that are not accessible to organisms, and so cannot be retained by them. How important are what I will call "demand-independent" element losses? Conversely, are there important pathways of element inputs that are "demand-dependent" (in addition to biological N fixation)? What are the implications of these pathways of element inputs and losses for the functioning of terrestrial ecosystems?

How do the cycles of different elements interact? Many analyses of nutrient cycling and limitation evaluate one element at a time, focusing for example on the dynamics of the N cycle; others consider multiple elements, but do so sequentially. However, organisms require a suite of elements simultaneously, at ratios that differ among organisms, and the processes that drive element inputs and output also affect multiple elements simultaneously. A recent analysis by Sterner and Elser (2002) expanded upon earlier work (Redfield 1958, Reiners 1986) to demonstrate that understanding the ratios of elements in organisms and ecosystems—their biological stoichiometries—can contribute substantially to understanding nutrient cycling and limitation. Can we integrate the biological stoichiometries of plants, animals, and decomposers with the geochemical stoichiometries of element inputs, outputs, and transformations, and develop a more fundamental analysis of biogeochemistry?

How do genotypes, species, and communities of organisms affect nutrient cycling and limitation in ecosystems? Many recent studies have evaluated the influences of particular species, functional groups, and/or levels of diversity on ecosystem productivity and nutrient retention (Tilman et al. 2001, Hooper et al. in press); other studies have demonstrated that although components of global change (e.g., warming, elevated CO_2) alter nutrient cycling directly, they have greater effects indirectly through their alteration of species composition (Hobbie 1995). One of the features that makes the Hawaiian Islands a useful model system is their low species diversity. I use this low diversity to evaluate the importance of population-level variation within species, and I use the consequences of biological invasions by species from outside Hawai'i to ask how nutrient cycling and limitation might differ in a more biologically diverse region.

These questions represent threads that run throughout this book, and I return to them in the last chapter for a more extended discussion. I do not answer all of these questions, or indeed any of them completely—rather I think research in Hawai'i has contributed to answering some of them, and to developing perspectives and informed speculations on the rest.

A brief road map for the book is as follows: chapter 2 begins by defining model systems, and describing how they are used in other fields. I briefly summarize the natural history of the Hawaiian Islands, and illustrate how Hawai'i has been used as a model system in analyses of evolution/ speciation, conservation biology, and culture as well as ecosystem studies. In chapter 3, I focus on the extraordinary environmental gradients within Hawai'i, describing how the combination of remarkable constancy in some of the environmental factors that control terrestrial ecosystems (sensu Jenny 1980), coupled with wide and well-defined gradients in other factors, makes the Islands useful for understanding nutrient cycling in particular. I conclude chapter 3 by describing a 4.1 million year substrate age gradient across Hawai'i that provides the focus for most of the remainder of the book. Chapter 4 describes patterns in productivity, nutrient availability, and nutrient cycling across that gradient, demonstrating that a biologically regulated positive feedback strongly reinforces the underlying geochemical controls of nutrient cycling as soils and ecosystems develop. In chapter 5, I summarize experimental studies of nutrient limitation across this substrate age gradient—evaluating which nutrients limit productivity in young, intermediate-age, and old sites, and with what consequences for nutrient cycling in ecosystems. Chapter 6 summarizes the inputs of elements to all of the sites along the age gradient, in the process demonstrating that long-distance transport of continental dust to Hawai'i contributes substantially to maintaining productivity and soil fertility there in the long term. In chapter 7, I evaluate the outputs of nutrients from sites along the gradient, and compare calculations of inputs versus outputs. These input-output budgets provide an independent check on estimates of inputs and outputs, and in the process demonstrate that there remains a lot about Hawaiian ecosystems that I do not understand. Finally, in chapter 8, I return to the questions raised above, and combine analyses based upon a simple simulation model with field results from Hawai'i and elsewhere to summarize what I think we know—and don't know—about nutrient cycling and limitation in terrestrial ecosystems.

Chapter Two

THE HAWAIIAN ISLANDS AS
A MODEL ECOSYSTEM

MODEL SYSTEMS have been widely used in molecular biology, where they have been responsible in part for that field's remarkable advances in the recent past. In ecology, model systems have received less emphasis. In this chapter, I define model systems, describe briefly how they are used in other fields, and distinguish model systems from microcosms and from well-studied systems, both of which are used commonly in ecology. I then review the natural history of Hawai'i, showing how the formation of the islands, their climates, and their isolation interact to create a unique—and uniquely useful—suite of ecosystems. Finally, I describe how the Hawaiian Islands have been used as a model system for the understanding of evolution and speciation, conservation biology, and human-land interactions.

Model Systems

A "model system" is a system—which could be a gene and its regulators, an organism, or an ecosystem—that displays a general process or property of interest, and does so in a way that makes it understandable. The model system is analyzed in order to understand the property or process, not the system itself. A model system can be useful because it is simpler than other systems of its type, so the property of interest is not obscured by others. It can be faster (in generation time, in turnover) than others, or smaller in size—both characteristics that facilitate experimentation. Finally, there might be idiosyncratic features of the particular system that make it useful for studying a particular process of general interest.

Krogh (1929) provided an early statement on the use of model systems in animal physiology: "For a large number of problems there will be some animal of choice, or a few such animals, on which it can be most conveniently studied" (Krogh 1929; in Krebs 1975).

Note that the focus is on animals that make a process ("problem") amenable to study; these animals are useful because they can provide understanding that is applicable to many or all animals. For example, research on the conduction of nerve impulses originally focused on the giant axons of the squid genus *Loligo* (Llinás 1999); these axons are more than a thousand times larger in diameter than most vertebrate nerve axons, and so could be studied directly with the electrodes available in the 1960s and 1970s. It was not the squid per se that was of interest, but rather nerve conduction in general, using the squid as a model system.

A more recent example is the nematode *Caenorhabditis elegans*. *C. elegans* is widely used for studies of the molecular biology of development because it is an integrated, multicellular organism, and yet it is remarkably simple. It has specialized "muscles," digestive tissue, a nervous system, and gametes; it undergoes embryogenesis and development and aging. At the same time, *C. elegans* is small (ca. one millimeter at maturity), has a short generation time, and is readily grown in culture. It is transparent, so that the movement and fate of individual cells and their daughters can be followed directly. Finally, an adult *C. elegans* is made up of only 959 somatic cells, whose formation and movements can be followed directly, manipulated experimentally, and understood on the basis of both genetics and biochemistry.

The use of *C. elegans* as a model organism was not based on the expectation that its developmental program would be just like that of other organisms; the greater complexity of most multicellular animals ensures otherwise. Rather, the expectation (now abundantly realized) was that the simplicity of the *C. elegans* system would allow many of the most fundamental mechanisms involved in development to be understood—and that the understanding gained by studying *C. elegans* would be useful in the analysis of more complex organisms (*C. elegans* Sequencing Consortium 1998). In this sense, *C. elegans* was selected as a model system because it represents a useful compromise between complexity and tractability. Other widely used model systems in physiology and molecular biology include *Drosophila melanogaster, Escherichia coli, Arabidopsis thaliana*, and the ubiquitous laboratory rats and mice.

Ecology also has made use of model systems, although not to the same extent as other fields. Perhaps the clearest expression is also the earliest I know of—Forbes' (1887) statement that a lake: "forms a little world within itself—a microcosm within which all of the elemental forces are at work and the play of life goes on in full, but on so small a scale as to bring it easily within the mental grasp."

Lakes have remained model systems for ecosystem studies because they are clearly bounded, well-mixed compared to terrestrial ecosystems, their primary producers and many consumers are small and short-lived, and they lack the persistent structures that characterize and complicate terrestrial ecosystems—and yet they are real, whole, persistent, functioning ecosystems. Aquatic ecologists have made the most of these advantages, providing the first description and/or the best mechanistic analysis of many fundamental ecosystem characteristics, including the inefficiency of energy flow in food chains (Lindemann 1942), top-down and bottom-up controls on ecosystem structure and functioning (Carpenter and Kitchell 1993), interactions between the population biology of individual species and the functioning of ecosystems (Frost et al. 1995), and the importance of stoichiometry in controlling nutrient cycling and other ecological interactions (Elser et al. 1996). These processes are important in all ecosystems; they were characterized first and/or best in lakes because of the tractability of that model system. (The alternative explanation, that aquatic ecologists are smarter than terrestrial ecologists, can be falsified readily by observation.)

Microcosms and Well-studied Systems

Model systems differ from microcosms and well-studied ecosystems—two related approaches that are more widely used in ecology. One of the essential factors of model systems, be they nematodes or plants or lakes, is that they are an integrated, functional, persistent example of the larger set of systems that they are meant to illuminate. They have realistic dynamics and controls because they are real. In this, they differ from microcosms, which are artificially constructed systems designed to include a subset of the organisms and interactions present in ecosystems. Microcosms are extraordinarily useful because they can be controlled, replicated, and used to focus research on particular interactions or processes of interest; species interactions in particular have long been studied in microcosms, with a precision that is unattainable in field studies (e.g., Gause 1934, Bohannan and Lenski 1997, Kerr et al. 2002). Microcosm-based studies of ecosystem processes have also been valuable (Bowden 1991, Bormann et al. 1993), but microcosms (unlike real systems) can yield misleading results because they fail to include organisms, properties, or processes that are fundamental to the functioning of real ecosystems (Carpenter 1996). This is not meant to diminish the value of microcosms, but rather to distinguish them clearly from model systems. The latter *are* ecosystems that have persisted for long periods of time.

Alternatively, systems may also be selected for study because a great deal already is known about them, and/or because other research is already in progress. Such well-studied systems provide a logical focus for additional research, because much of the background and context for evaluation of a particular property or process is available. Model systems and well-studied systems intergrade, because a model system that provides useful insight in one area may become widely used and relatively well-known. For example, the Hubbard Brook Experimental Forest initially was selected for ecological study because its watertight bedrock allowed the direct calculation of hydrologic nutrient budgets (e.g., Likens et al. 1977)—in other words, because it was a model system. However, the context provided by those budgets and by the presence of many active researchers has made Hubbard Brook into a site that attracts research in large part because so much is known already about its dynamics—because it is now a well-studied system. Conceptually, it's a well-studied system if work is located there primarily to build upon other research; it's a model system if work is located there primarily because characteristics of the system itself make it useful to answering a general question.

A BRIEF NATURAL HISTORY

For many ecological questions, oceanic islands in general—and the Hawaiian Islands in particular—provide researchers with a useful compromise between complexity and tractability, at least as well as does *C. elegans*. Here I summarize the natural history of the Hawaiian Islands, to provide a background for understanding their usefulness for a wide range of research. More detailed information can be found in Carlquist (1980), MacDonald et al. (1983), Juvik et al. (1998), Mueller-Dombois and Fosberg (1998), and Ziegler (2002).

The Formation of the Hawaiian Islands

VOLCANISM

The Hawaiian Islands are being created by a plume of hot, buoyant magma that originates deep in Earth's mantle and pushes through the relatively thin oceanic crust. This plume, or "hot-spot," has been active for at least 80 million years—and were it not for plate tectonics, it could have built a mountain to rival Olympus Mons on Mars, the largest volcano in our solar system. However, the plume reaches the surface within the Pacific tectonic plate, which is sliding northwest at a rate of about 8–9 centimeters per year. Instead of a single massive mountain, it has created a chain of volcanic islands, atolls, and submerged seamounts that stretches more than 6000 km from Hawai'i to the Kurile Trench near Kamchatka (fig. 2.1) (Clague and Dalrymple 1987). Many other hot-spots are known, and many island archipelagoes have originated from them, but the Hawaiian hot-spot and archipelago is the most spectacular on Earth.

The hot-spot is now centered near the southeastern edge of the Islands, where it feeds magma to two very active volcanoes on the Island of Hawai'i (Mauna Loa and Kīlauea) and a rapidly growing submarine volcano farther to the southeast (Lō'ihi). Two other volcanoes (Hualālai on Hawai'i and

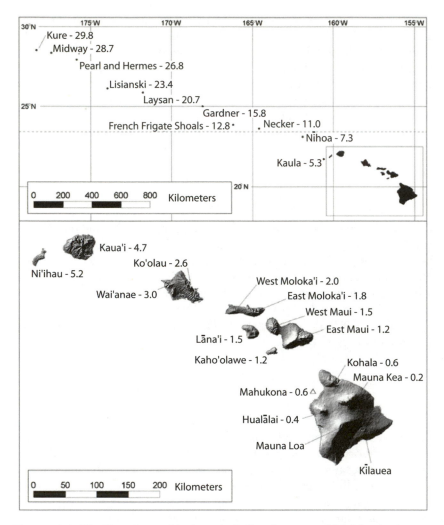

FIGURE 2.1. The Hawaiian archipelago, including the current high islands, the ages of their major volcanoes in millions of years, and atolls. Modified from Price (2002).

Haleakalā on Maui) have been active in the past 250 years. Moving northwest away from the current location of the hot-spot, individual volcanoes are progressively older, both within and between islands (fig. 2.1).

The frequency, accessibility, and relative simplicity of Hawaiian volcanoes has attracted a great deal of volcanological research (the islands are a model system for understanding hot-spots and their archipelagoes), and the life history of an individual Hawaiian volcano is well characterized.

FIGURE 2.2. Mauna Loa Volcano (4168 m), photographed across the Kīlauea caldera on the island of Hawai'i, reproduced with permission from MacDonald et al. (1983). The characteristic shield shape reflects a history of frequent eruptions by relatively fluid lava flows.

A rapidly growing young shield volcano is fed by the rising plume of magma from the hot-spot, yielding frequent eruptions of hot, relatively fluid lava. At this stage the lava produced is tholeiitic basalt; its chemistry reflects that of the mantle plume, with some variation in magnesium concentrations caused by variable crystallization of olivine (fractionation) in the volcanoes' plumbing. Fast-moving lava flows can travel far from their vents in the summit or in rift zones on the volcanoes' flanks, and so each volcano develops a gentle, more or less continuous slope from sea level to its summit. Young Hawaiian volcanoes are exemplars of shield volcanoes, built up of lava flows piled upon lava flows with a shallow angle of repose (fig. 2.2). In contrast, the stratovolcanoes that arise at continental margins where tectonic plates are subducted beneath others produce more viscous lava, more explosive eruptions, and steeper slopes—as the contrast between the profile of the 4168 meter shield volcano Mauna Loa and that of the 4393 m stratovolcano Mt. Rainier illustrates (fig. 2.3).

FIGURE 2.3. The profile of the 4168 m shield volcano Mauna Loa relative to that of the 4383 m stratovolcano volcano Mt. Rainier. This comparison understates the relative mass of Mauna Loa, which extends to its base an additional ~6000 m below sea level.

More than 95 percent of the mass of most Hawaiian volcanoes is tholeiitic basalt produced during this shield-building phase. Eventually, however, the movement of the Pacific plate transports a volcano northwest away from the hot-spot. Eruptive activity then slows, and the magma that does reach the surface spends longer in reservoirs within the volcano. There, it partially crystallizes and fractionates; the material that erupts is more alkalic (richer in calcium) than during the shield-building stage (Wright and Helz 1987). Most older Hawaiian volcanoes are covered with a cap of alkalic lavas up to hundreds of meters thick, produced by late-stage volcanism.

After tens to hundreds of thousands of years of late-stage volcanism, each volcano loses contact with the plume, and eruptive activity ceases. Still later, there can be a rejuvenation of activity as the continued movement of the Pacific plate carries the eroding volcano over the edge of a depression in the crust surrounding the hot-spot. This rejuvenation-stage or post-erosional volcanism produces lava even more alkalic than late-stage volcanism. Rejuvenation-stage eruptions are infrequent, but they have built some of the most familiar features of the Hawaiian landscape, including Diamond Head on O'ahu.

SUBSIDENCE, SLIDES, AND EROSION

In opposition to the volcanism that builds the islands, a number of processes combine to bring them down. Quantitatively, the most important of these is subsidence. The mass of a growing volcano depresses the relatively thin oceanic crust beneath it, creating a doughnut-shaped depression. The youngest island, Hawai'i, now is subsiding at a rate of ~2.5 mm/yr (Ludwig et al. 1991, Moore and Clague 1992); Maui, the next-youngest, has subsided ~1500 m in 1.2 million years (mostly during and immediately after the shield-building stage) (Clague 1996, Price and Clague 2002).

A second mechanism of mass loss is slumps and slides from a volcano's unsecured flanks (the sides that are not buttressed by adjoining volcanoes

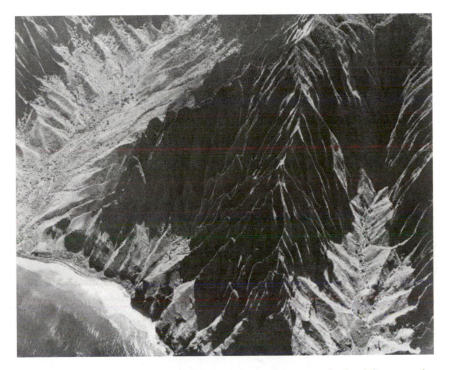

FIGURE 2.4. The consequences of millions of years of erosion for land-forms on the north side of the island of Kauaʻi. Reproduced with permission from MacDonald et al. (1983). Between 4 and 4.5 million years ago, the island of Kauaʻi was a shield volcano (or two; see Holcomb et al. 1997) much like Mauna Loa is now.

on the same island). These slides rapidly remove up to 5000 km^3 of material; they can give rise to spectacular tsunamis (Moore et al. 1994). One such slide or set of slides removed the northeastern side of the Koʻolau Volcano on the Island of Oʻahu, opposite Honolulu, to a point beyond the summit. The more familiar processes of chemical weathering and fluvial erosion also sculpt the landscape of the older islands, creating spectacular valleys and cliffs (MacDonald et al. 1983) (fig. 2.4).

This combination of creative and destructive forces yields a progressive change in land forms across the islands, from smooth, active shield volcanoes at the southeastern margin to increasingly eroded older volcanoes to the northwest (fig. 2.5). The Hawaiians long recognized the progressive nature of their islands; their traditions speak of the volcano goddess Pele traveling from island to island from northwest to southeast, kindling the activity of new volcanoes and causing the extinction of older ones as she moved on.

FIGURE 2.5. Erosion and the development of land forms across the Hawaiian Islands, prepared by Stephen Porder and modified from Vitousek et al. (in press). (a) The active Kīlauea volcano displays little influence of fluvial erosion; its topography results from volcanic and tectonic activity (craters, cinder cones, block faults on the south flank). (b) Kohala Mountain, which was built ~500 ky ago (ky = 1000 yrs), maintains most of its shield surface and many cinder cones; however, the windward (northeast) side has been altered by massive slides and dissected by numerous stream valleys up to several hundred meters deep. (c) West Maui Mountain (1500 ky) has been carved by erosion; intact shield surfaces can be found mainly on the dry southwestern side of the mountain. (d) Kaua'i (~4500 ky) has been dissected even more thoroughly than West Maui; only pockets of shield topography remain, as discussed in chapter 3.

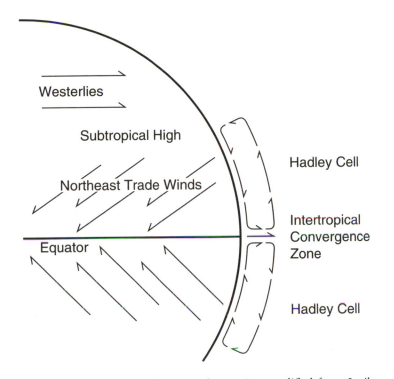

FIGURE 2.6. Atmospheric circulation in the tropics, modified from Juvik et al. (1998). Most of the time, the Hawaiian Islands lie within the northern hemisphere Hadley cell, under the influence of the northeast trade winds.

Once a volcano erodes and subsides to sea level, one further building process takes place. Fringing coral reefs coalesce and build a coral atoll on the platform of the vanished volcano. Such atolls can persist tens of millions of years after the volcano that supports them disappears below the surface.

Determinants of Climate

The regional climate of the Hawaiian Islands is controlled by their location in the northeast trade wind zone, just south of the Tropic of Cancer, and by the ocean that surrounds them. The trade winds are the surface expression of the tropical Hadley circulation cells, within which solar-heated air rises off the surface near the equator, flows north and south at high altitude, descends near 30°N (making the subtropical Pacific high pressure area) and flows back towards the equator near the surface (fig. 2.6). In the northern hemisphere, this surface flow appears to be deflected towards the right by the Coriolis effect, giving rise to the northeast trade winds.

The heated air that rises in the intertropical convergence zone (where the northeast and southeast trade winds come together near the equator) loses moisture as it ascends, and so has very low humidity where it descends to the surface at 30°. This air picks up moisture from the ocean as it flows back towards the equator in a relatively thin boundary layer, from the surface up to an altitude of ~2000 m. A temperature inversion marks the boundary between moist air near the surface and the very dry air above (Giambelluca and Schroeder 1998).

Because Earth's axis is tilted 23.5° relative to plane of its orbit, the latitude where Earth is heated most—and so the Hadley cells—shift north in the northern hemisphere summer and south in the winter. Hawai'i almost always lies within the trade wind zone in the northern hemisphere summer, but the islands often come under the influence of the subtropical Pacific high and of temperate low pressure areas in the winter.

Sea-level temperature in Hawai'i is controlled by its maritime location; the massive heat capacity of the surrounding ocean buffers the seasonal range in mean monthly temperature to no more than 4°C. Even daily extremes are muted; the lowest temperature ever recorded in the coastal city of Hilo on the Island of Hawai'i was 11°C, while the highest temperature ever was 32°C. Mean annual temperature decreases with increasing elevation at an environmental lapse rate of ~6.4°C/1000 m, up to the trade wind inversion at ~2000 m. Above the inversion, mean annual temperature decreases more slowly, reaching 4°C at the 4205 summit of Mauna Kea (Giambelluca and Schroeder 1998)—a name meaning "White Mountain" in recognition of its frequent winter snow cover.

Precipitation over the open ocean near Hawai'i averages 600–800 mm/yr. However, the Hawaiian Islands are large enough to shape their own precipitation. The trade winds are forced upwards when they encounter the northeast side of Hawaiian mountains, cooling as they rise and thereby causing condensation that brings heavy rain to these windward slopes. Extensive areas receive precipitation in excess of 4000 mm/yr, plus a poorly known but substantial input of intercepted cloudwater (Giambelluca et al. 1986). One site, Wai'ale'ale on the Island of Kaua'i, averages > 11,000 mm/yr, making it a contender for the rainiest place on Earth. On Hawaiian mountains that do not reach above the tradewind inversion (~2000 m), the wind is stripped of moisture as it passes over the summit and descends to the leeward coast. These leeward areas are in a rain shadow, with frequent sunshine and little precipitation. Large areas receive less than 500 mm/yr, and Kawaihae on the Island of Hawai'i receives < 200 mm/yr (fig. 2.7) (Giambelluca et al. 1986). It is no coincidence that most resort hotels are clustered on leeward coasts.

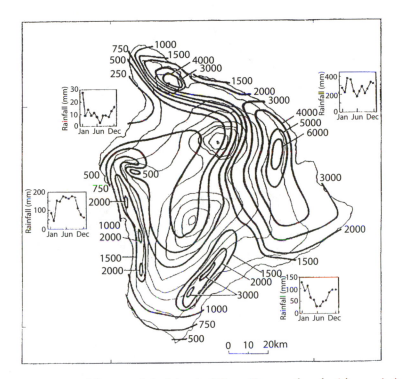

FIGURE 2.7. Rainfall map of the Island of Hawai'i, reproduced with permission from Vitousek (1995) and based on information in Giambelluca et al. (1986). Fine lines are 500 m elevation contours; the coarse lines are rainfall isohyets. The inset maps show the seasonality of precipitation in different areas of the island; Hilo on the northeast is wet year round, Kawaihae on the northwest is dry most of the year, Kainaliu on the southwest receives summer rain, and Pāhala on the southeast receives winter rain.

The pattern of precipitation on larger, higher volcanoes is more complex. The summit regions of mountains that reach into the dry air above the tradewind inversion generally receive < 500 mm/yr of precipitation (fig. 2.7). Moreover, the trade winds flow around, not over, the larger mountains—leading to wind-sheltered rather than rain-shadowed zones on their leeward slopes. The absence of zonal winds allows a diurnal land-sea breeze cycle to dominate the local climate of these regions. This cycle is best developed on the Kona coast of the Island of Hawai'i, in the lee of Mauna Loa and Hualālai volcanoes. Diel heating of the land causes air to rise above the island, pulling in moisture-laden onshore breezes that cool as they rise along the mountain slopes, causing afternoon cloudiness and rain (fig. 2.8). At night, the land cools faster than the ocean, driving

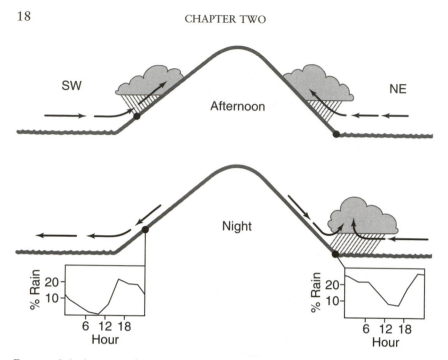

FIGURE 2.8. Sources of precipitation on largest Hawaiian volcanoes, modified from Carlquist (1980). Daily heating of the land surface causes air to rise above the volcano, drawing moist air in off the sea and bringing afternoon rain to what otherwise would be dry leeward slopes. Note the afternoon peak in hourly precipitation in Kona (left), on the southwest flank of Mauna Loa (Giambelluca and Schroeder 1998). On windward slopes, cooling of the land at night causes downslope winds that converge with the trade winds, bringing night rains to the windward coast (as shown on the right of the inset map of hourly precipitation in Hilo).

downslope winds of cold dense air that drive the clouds out to sea. This cycle is strongest in the summer, when the sun is directly overhead at midday—and so the montane rainforests of Kona receive most of their rainfall on summer afternoons. Downslope night breezes also affect the northeast (windward) areas of the largest mountains. Here, the convergence of onshore tradewinds with the cold downslope wind at night bring predawn rainfall to the windward coast of the Island of Hawai'i (and to a lesser extent, Maui)—making Hilo the rainiest city in the United States (fig. 2.7, 2.8).

When the intertropical convergence and associated Hadley cells are displaced to the south during the northern hemisphere winter, temperate-zone low-pressure systems that are preceded by southeasterly winds and followed by westerlies occasionally cross the Islands, bringing widespread rain. Leeward areas receive most of their annual precipitation from such systems, and so have winter precipitation maxima (fig. 2.7) (Giambelluca and Schroeder 1998). Infrequent hurricanes also can bring widespread rains.

It is worth noting that although the climate of Hawai'i is strongly buffered by the ocean, it is far from constant. On short time scales, the El Nino phase of the El Nino southern oscillation (ENSO) system can bring severe droughts to the islands (Loope and Giambelluca 1998). On longer time scales, paleoecological studies demonstrate that what are now wet windward forests were somewhat cooler and substantially drier during the last glacial maximum 21,000 years ago (Hotchkiss 1998, Hotchkiss and Juvik 1999), and an ice cap then covered the summit of Mauna Kea (Porter 1979).

Isolation

The Hawaiian Islands are the most isolated archipelago on Earth, in that they are farther from any continent or other high islands than is any other archipelago. This isolation strongly shapes their biology and human history, and their usefulness as a model system for ecological studies.

Biologically, oceanic islands are defined by having rates of speciation that are greater than rates of colonization, with the consequence that many species are endemic (found nowhere else on Earth). Hawai'i fits this definition; over 90 percent of the native vascular plants are endemic, along with all of the native passerine birds and most of the insects (Wagner et al. 1990, Loope 1998, Mueller-Dombois and Fosberg 1998). The few species that did find their way to Hawai'i naturally, whether carried by the jet stream from Asia or in mud on the feet of migratory birds (Carlquist 1982, Juvik 1998), occupy an extraordinarily broad range of environments. Some remain a single widespread species; the most extreme of these is the dominant tree *Metrosideros polymorpha* ('ōhi'a) in the Myrtaceae (Dawson and Stemmermann 1990). Other early colonists have speciated extensively in Hawai'i, as discussed later in this chapter. At the same time, some major groups of organisms never made it to Hawai'i naturally, including ants, snakes, and terrestrial mammals (other than a single insectivorous bat). Consequently, the biota is "disharmonic;" it differs in kind from the biota of continents, either lacking some major functional groups of organisms or having unusual species that partly fill a functional role that is carried out by very different organisms on continents (e.g., in Hawai'i and other islands, now-extinct endemic flightless birds filled something approximating the role of browsing mammals; James 1995, Steadman 1995).

The isolation of the Hawaiian Islands also means that people arrived very late, only 1200–1500 years ago (Kirch 1985). The Polynesian discoverers were on voyages of colonization; they brought with them a suite of plants, animals, agricultural strategies, and social structures. On arrival, they quickly made up for the long absence of humans, developing several forms of intensive agriculture, high population densities, and complex societies (Kirch 1985, 2000, Cordy 2000). They and their domestic animals also hunted some of the native species to extinction, and cleared and transformed much of the arable land (Kirch 1985, Burney et al. 2001). Later human arrivals, beginning in the 1770s, expanded the area of human-dominated habitat and introduced a stunning variety of plant and animal species, and of diseases. Nevertheless, compared to most of Earth, the history of human occupancy of the land is short.

EVOLUTION, CONSERVATION, AND CULTURE

Although most of this book focuses on ecosystem-level nutrient cycling, the natural history of the Hawaiian Islands makes them useful as a model system for a broad variety of questions in evolution, ecology, and anthropology. I discuss these areas briefly here, and then introduce analyses of ecosystem dynamics in the next chapter. I believe that the breadth of areas for which Hawai'i represents a useful model system multiplies its potential value; not only can the islands serve as a model within a number of fields of research, but they can facilitate synthesis across ecosystem ecology, evolution, conservation biology, anthropology, and other fields.

Evolution and Speciation

The contribution of islands to our understanding of evolution and speciation has long been clear—at least since Darwin (1845) in the Galapagos felt "near, both in space and time, to that great fact, that mystery of mysteries, the first appearance of new beings upon the Earth." What Darwin saw—and what many have since exploited—is that the relationships among species within archipelagoes are clearer than those on continents, with recognizably similar but nevertheless distinct species occupying different islands in an archipelago, and sometimes different habitats within an island. Both the isolation of islands and their relative simplicity combine to make the analysis of speciation relatively straightforward.

Most oceanic archipelagoes are useful for understanding aspects of evolution and speciation—and thanks to Darwin's visit and subsequent work, the Galapagos can claim primacy in this area (Grant 1999). However, neither the Galapagos nor any other archipelago can compare to the Hawaiian Islands in number of endemic species, their distinctiveness, or the range of environmental variation across which they have radiated. Among the most spectacular of these radiations are the Campanulaceae

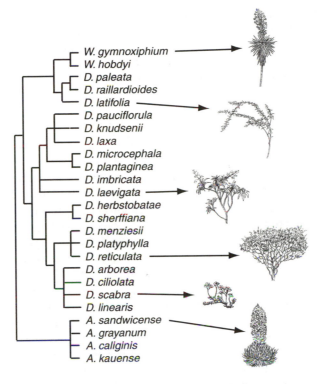

FIGURE 2.9. Evolutionary radiation of the silversword alliance (Asteraceae) in the Hawaiian Islands, modified from Baldwin (1997). An annual tarweed from California has radiated into three genera (*Argyroxyphyllum, Dubautia,* and *Wilkesia*) in the Hawaiian Islands; species that have evolved here include herbs, giant rosette plants, shrubs, lianas, and trees.

(lobeliads) with 125 species in six genera, descended from a single introduction (Givnish et al. 1995); the silversword complex (30 species in three genera, including trees, shrubs, herbs, and lianas; all descended from an annual tarweed, see fig. 2.9) (Baldwin and Sanderson 1998); the drepanid honeycreepers, with an extraordinarily diverse range of sizes, bill morphologies, feeding strategies, and vibrant colors (Fleischer and McIntosh 2001); and the drosophilid fruit flies with > 500 species, perhaps half as many as in the rest of the world combined (Kaneshiro 1995). Many of these radiations have given rise to closely related species in widely contrasting environments; for example the silversword complex includes species adapted to dry alpine conditions and very wet rainforests. These evolutionary radiations make Hawai'i an unparalleled natural laboratory— a model system—for understanding evolution and speciation. Evolutionary

biologists have made abundant use of this resource to understand fundamental evolutionary processes as diverse as mechanisms of speciation
(Carson 1986), the importance of phylogenetic constraints to organismal
physiology in differing environments (Robichaux et al. 1990), and the
evolution and maintenance of genome size (Petrov et al. 2000).

Conservation Biology

The isolation that fosters spectacular evolutionary radiations leaves the
products of those radiations extraordinarily vulnerable to human activities,
especially the introduction of plants, animals, and microbes from around
the world (Elton 1958, Carlquist 1974). Consequently, island biotas have
been among the first victims of the modern anthropogenic mass extinction
(Cuddihy and Stone 1990, Steadman 1995). Many species have been lost,
and many more are on the verge of extinction; Hawaiian species make up
nearly 30 percent of recognized threatened and endangered species in the
United States, although the Hawaiian Islands represent only ~0.2 percent
of the land area of the United States.

The widespread extinction and endangerment of island species is devas-
tating culturally, aesthetically, and scientifically. Nevertheless, this concentration of endangered species gives conservation biologists the opportunity
to learn how to recover multiple species on the brink of extinction (Loope
1998). Not all such efforts at recovery will be successful—but we can learn
from the inevitable failures as well as successes, and apply much of what
we learn to endangered species and biotas around the world.

Cultural Evolution

Oceanic islands are useful for understanding the development of human
societies—their cultural evolution—for many of the same reasons as for
biological evolution. For most of their history, human societies on oceanic
islands were more or less isolated from other human societies—rarely completely, but more so than was true on any continent. Consequently, outside influences and interactions were less important and local influences
relatively more important to the development of their cultures. Islands
differ from each other in fundamental ways—including size, complexity,
climate, and soil fertility as well as attributes of their founding cultures—
and these differences interact with social processes to shape their societies. Moreover, many island cultures have left physical and linguistic
records of their development over time, records that can be interpreted
directly in terms of divergence from common origins (Kirch 2000, Kirch
and Green 2001).

These features reach their fullest development in the islands of Polynesia. The Polynesians are a well-defined people and culture who colonized an extraordinarily broad range of islands fairly recently; many of their islands are among the most isolated inhabited lands on Earth. They carried a suite of crops and domestic animals along on their voyages, and practiced intensive agriculture using a wide variety of cropping systems that were adjusted to local conditions (Kirch 1994, 2000). On many islands (including Hawai'i), they supported large populations in societies with a high degree of social and cultural complexity—societies that evolved along different paths, in the different islands of Polynesia. Within the Hawaiian Islands as elsewhere in Polynesia, a strong and repeatable contrast developed between both the production systems and the social structures of societies based on dryland farming and those based on wetland agriculture irrigated by the surface streams (Kirch 1994). Here the land helped to shape human culture, in ways that can be documented in Polynesia but that may apply to continental as well as island societies.

Just as land influenced Polynesian societies, Polynesian societies influenced land. In many places, intensive agriculture enhanced soil erosion, decreased soil fertility, and caused the extinction of many species—thereby threatening the sustainability of Polynesian societies themselves (Kirch 1997, 2000). Those societies then faced the challenge of making a transition from intensive, exploitative use of their island's obviously limited resources, to more sustainable use of those resources. Polynesian histories contain dramatic examples of the social and environmental consequences of failing to make that transition, but they also include impressive successes (Kirch 1997). We ourselves now face the challenge of a global transition to sustainability (National Research Council 1999); can we learn from societies that succeeded or failed on smaller worlds?

Chapter Three

GRADIENTS IN ENVIRONMENTAL FACTORS, GRADIENTS IN ECOSYSTEMS

THIS CHAPTER makes use of the model system concept and of the natural history of the Hawaiian Islands, both described in chapter 2, and applies them to understanding the structure and functioning of ecosystems. Because of their wide variation in climate and geological age, the Hawaiian Islands support an extraordinary variety of ecosystems in a very small geographical area; these ecosystems range from deserts to cloud forests, alpine systems to steamy lowland rainforests, fresh hot rock to ancient deeply leached soils, and much more. Nevertheless, many of the potential controls of ecosystem structure and functioning can be held constant in Hawai'i, to an extent that cannot be matched elsewhere. This combination of substantial (and well-defined) variation in ecosystems and unusual simplicity in many of their controlling factors makes Hawai'i a useful model system for understanding many of the fundamental mechanisms that control the structure and functioning of terrestrial ecosystems.

The State Factor Framework

The "ecosystem state factors" defined by the soil scientist Hans Jenny (1941, 1980)—climate, organisms, relief or topography, parent material, and time—provide a useful framework for analyzing patterns and probable causes of variation in ecosystems. Three of these factors (parent material, relief, and organisms) can be held constant in the Hawaiian Islands, to an extent that can only be dreamed of in most continental ecosystems (Vitousek 1995, Vitousek et al. 1995a, Vitousek 2002). As described in the previous chapter, the *parent material* of the Hawaiian Islands is almost all basaltic rock derived from the Hawaiian hot spot. Although the texture of this material differs substantially, from massive pāhoehoe lava to jumbled 'a'ā to volcanic ash (fig. 3.1), its chemistry is reasonably constant in both space and time (Wright 1971). Late-stage alkalic volcanism differs chemically from the shield-building theoleiitic basalts—but against the spectrum of rocks found in most regions, these differences are subtle. All Hawaiian ecosystems begin their development in rock that differs relatively little in chemistry.

FIGURE 3.1. Pāhoehoe versus ʻaʻā lava, reproduced with permission from Mac-Donald et al. (1983). Pāhoehoe (left) has a massive pavement-like surface with frequent cracks; the surface of ʻaʻā is a jumble of rocks of various sizes, with a solid core.

Eruptions of the active young volcanoes produce frequent, relatively fluid lava flows that build shield volcanoes nearly lacking in coarse-scale *relief* (fig. 2.2). Once eruptive activity slows, fluvial erosion and more massive landslides create spectacular topography (fig. 2.4). However, areas of intact constructional surfaces of the original shield volcano can be identified even on > 4 million-year-old Kauaʻi, the oldest of the high islands.

The isolation of the Hawaiian Islands greatly reduces their diversity relative to continental areas, and thereby helps to keep the *organism* factor simple. The myrtaceous tree *Metrosideros polymorpha* (locally "ʻōhiʻa") in particular is found (often as the dominant) in sites from sea level to treeline, as the first woody plant colonizing young lava flows to the oldest high islands, and from sites receiving < 400 mm to > 10,000 mm of annual precipitation (Dawson and Stemmermann 1990). *Metrosideros* thus provides a consistent biotic background across a very wide range of conditions. Moreover, recent biological invasions represent a fundamental change in the organism factor, one that can be used to understand how individual species shape the properties of ecosystems (Vitousek and Walker 1989, Mack et al. 2001).

Although parent material, relief, and organisms are nearly constant (or can be held constant) across Hawaiʻi, *climate* and *time* vary enormously.

For *climate,* temperature and precipitation vary substantially and largely orthogonally (fig. 2.7). Temperature decreases with increasing elevation, in Hawai'i as elsewhere, while below the tradewind inversion precipitation is controlled primarily by slope aspect relative to the northeast trade winds. Rainfall gradients are unusually steep—windward slopes on the north and east side of most islands receive > 4000 mm/yr, while leeward areas < 15 km away receive 10% or less of that precipitation (Giambelluca et al. 1986, Giambelluca and Schroeder 1998). Overall, the Islands support an extraordinarily large and well-characterized range of climatic conditions in a very small geographical area (George et al. 1987).

The age of the substrate in which ecosystems develop varies on two *time* scales within Hawai'i. On the active volcanoes, individual lava flows create a matrix of young and old substrates that have been precisely mapped and dated (Lockwood 1995, Wolfe and Morris 1996). The frequency of eruptions ensures that most substrates on the active young volcanoes are less than 4000 years old, making these systems useful for evaluating primary succession and early stages of soil development. On a longer time scale, the movement of the Pacific tectonic plate across the Hawaiian hot-spot creates a linear array of volcanoes and islands (fig. 2.1) that stretches from active volcanoes to > 4 million-year-old high islands, and to a much older chain of atolls (Carson and Clague, 1995). This array of volcanoes can be used to understand controls of long-term soil and ecosystem development.

Environmental Gradients as Model Systems

My fundamental approach to understanding ecosystem structure and functioning in Hawai'i has been to define gradients of sites that differ in one controlling factor (e.g., substrate age) but are as alike as possible in others, and to use these gradients as a starting point for process measurements, integration, experiments, and models. This approach is widespread, for example in chronosequence studies of succession (Pickett 1989, Chapin et al. 1994, Lichter 1998) and toposequence studies of soils (Schimel et al. 1985), but it is particularly useful in Hawai'i because "as alike as possible" is more alike than can be achieved elsewhere, and because the factor of interest can be varied over such a broad range.

I recognize that the factors underlying these environmental gradients are coarse, ultimate controls of ecosystems, and there can be several layers of mechanisms and feedbacks between them and proximate controls of ecosystem properties and processes (e.g., Robertson 1989). Consequently, merely defining a gradient and observing how an ecosystem property varies along it can yield only limited insight. However, gradients make wonderful starting points for dynamic process-level measurements, for integrations based on stable isotopes, for experiments that analyze controls of ecosystem functioning, and for developing and testing simulation models. Through these approaches, proximate and ultimate controls can be identified and their interactions and feedbacks elucidated with more power than is available to detailed analyses of a single site, or to comparative analyses of several unrelated sites.

Researchers have made use of environmental gradients in Hawai'i for many years. Early primary succession was characterized by Forbes (1912), MacCaughey (1917), Skottsberg (1941), Atkinson (1970), and Eggler (1971); Sherman and Ikawa (1968) and Yost et al. (1982) evaluated soil sequences. The Island Ecosystems project of the International Biosphere Program utilized an elevational gradient (Mueller-Dombois et al. 1981), and studies of forest dieback (Mueller-Dombois 1986, 1992) and crop growth (George et al. 1987) have similarly used environmental gradients.

Most of this book builds upon analyses of ecosystem properties and processes distributed along a > 4 million year gradient in substrate age across the Islands. Studies along that gradient speak directly to understanding nutrient cycling and the causes and consequences of nutrient limitation in terrestrial ecosystems. Before describing that gradient, however, I will introduce several other major gradients in Hawai'i.

Temperature

An individual lava flow represents an ecologist's dream of an elevational transect—it is a linear system underlain by the same substrate, formed at the same time, often reaching from above treeline to the sea (fig. 3.2). Below the trade wind inversion on the windward side of the island of Hawai'i, mean annual temperature decreases by nearly 13°C from sea level to 2000 m, and precipitation is > 2000 mm/yr everywhere (Juvik and Nullet 1994). Even the vegetation is dominated by the same species throughout (Aplet and Vitousek 1994). Measurements of net primary production and litter decomposition on such flows (Vitousek et al. 1994, Raich et al. 1997, Russell et al. 1998) show that net primary production (NPP) increases linearly and litter decomposition increases exponentially in response to increasing temperature (fig. 3.3), resulting in progressively smaller equilibrium pools of soil C in warmer, low-elevation sites (Raich et al. 2000). By measuring changes in C isotopes following forest-to-pasture conversion on an older Mauna Kea lava flow, and using models calibrated with differences in ^{13}C and ^{14}C in soils along the gradient, Alan Townsend demonstrated that slowly cycling pools of soil organic matter (SOM) also decompose exponentially more rapidly in progressively warmer sites (fig. 3.4) (Townsend et al. 1995, 1997). These results provide a long-term complement to short-term experimental studies of the effects of

FIGURE 3.2. The 1859 lava flow at the left-center of this satellite image/digital elevation model stretches from well above treeline to the sea on the northwest flank of Mauna Loa volcano. The image was provided by Oliver Chadwick. This and other lava flows represent magnificent transects across a broad range of temperature and elevation.

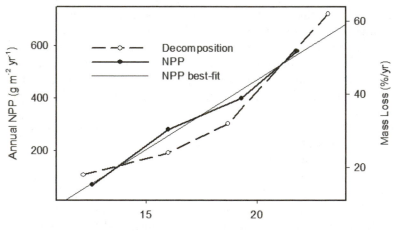

FIGURE 3.3. Above-ground net primary production (NPP; Raich et al. 1997) and the decomposition of *Metrosideros polymorpha* leaf litter (Vitousek et al. 1994) as a function of mean annual temperature along an elevational gradient on the windward 1855 lava flow, Mauna Loa. Across this flow, NPP increases linearly while decomposition increases exponentially with increasing temperature.

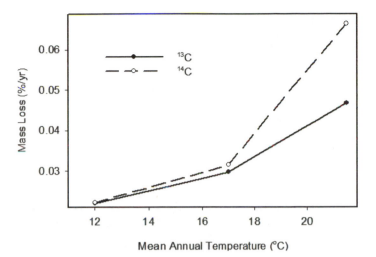

FIGURE 3.4. Turnover of the slow pool of soil organic matter as function of temperature along an elevational gradient on Mauna Kea Volcano, from Townsend et al. (1995). Solid symbols represent calculations based on ^{13}C turnover, following replacement of forest vegetation with the C_3 photosynthetic pathway by C_4 pasture grasses; hollow symbols are based on dilution of the ^{14}C spike from atmospheric testing of nuclear weapons. Decomposition of soil organic matter as well as litter (fig. 3.3) increases exponentially with increasing temperature.

warming on soil and ecosystem carbon balances (Peterjohn et al. 1994, Harte and Shaw 1995, Rustad et al. 2001, Melillo et al. 2002); they suggest that the equilibrium response of soils to warming temperatures is likely to be a net loss of soil C to the atmosphere.

Precipitation

Hawai'i supports some of the most dramatic precipitation gradients on Earth (fig. 2.7). Water/nutrient interactions have been evaluated on a number of these, including (1) a set of sites receiving 500 mm/yr to 5500 mm/yr, on ~2500 yr old substrate on the island of Hawai'i (Austin and Vitousek 1998, 2000; Austin 2002); (2) sites receiving from < 200–3000 mm/yr, on ~150,000 yr old substate on Hawai'i (Kelly et al. 1998, Chadwick and Chorover 2001, Chadwick et al. 2003); (3) a gradient from 2200–5000 mm/yr on Maui (Schuur et al. 2001); and (4) *Acacia koa*-dominated sites receiving from 500–2000 mm/yr on ~4,000,000 yr old substrate on Kaua'i (Harrington et al. 1995).

Among other things, we determined the natural abundance of ^{15}N in plants and/or soils along several of these gradients. The natural abundance of ^{15}N in ecosystems reflects integrated losses of N by fractionating

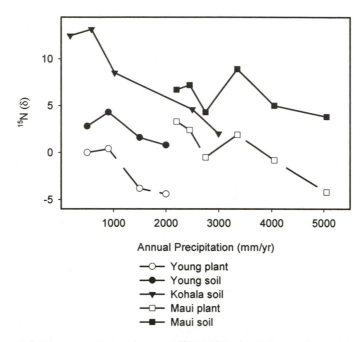

FIGURE 3.5. The natural abundance of ^{15}N (δ^{15}N, the difference between samples and an atmospheric standard, in parts per thousand) in soils and plants along three precipitation gradients in the Hawaiian Islands, including young sites on Mauna Loa and Hualālai analyzed by Austin and Vitousek (1998), Kohala sites from Uebersax (1996), and Maui sites from Schuur et al. (2001). Both older and drier sites are relatively enriched in ^{15}N.

pathways (Högberg and Johannison 1993)—and so the extent of ^{15}N enrichment can be used as an index of the openness of N cycling. Along Hawaiian rainfall gradients, dry sites are systematically enriched in ^{15}N relative to wet sites (fig. 3.5). The values for ^{15}N observed along any of these precipitation gradients depends upon substrate age as well as on precipitation, for reasons that will be discussed later, but plants and soil are relatively enriched in ^{15}N in drier sites at any given substrate age. The same pattern has been observed globally (Handley et al. 1999).

The Mauna Loa Matrix

The active shield volcano Mauna Loa on the Island of Hawai'i covers 5500 km^2, more than a third of the total area of the Hawaiian Islands, and reaches an elevation of 4168 m (fig. 2.2). Sites on Mauna Loa range in mean annual temperature from 5° to 24°C, in mean annual precipitation from ~250 mm to > 6000 mm/yr, and in surface age from < 20 years to tens of thousands of years. To evaluate how this spectacular variation controls the development and dynamics of terrestrial ecosystems, we have made use of a set of lava flows that reach from above tree line towards the sea on the wet northeast and the dry northwest flanks of Mauna Loa (fig. 3.6). On each flank, we selected a young (< 150 yr) and an old (> 3000 yr) flow of both the 'a'ā and pāhoehoe textures (fig. 3.1). This matrix provides sites that differ widely and largely independently in temperature, precipitation, substrate age, and parent material texture—yet are virtually identical in parent material chemistry, topography, and the dominance of *Metrosideros polymorpha* (Vitousek et al. 1992).

The Mauna Loa matrix has been used for a wide range of studies, including net primary production and nutrient cycling (Raich et al. 1997), decomposition (Vitousek et al. 1994, Russell et al. 1998), vegetation composition and biomass (Aplet and Vitousek 1994, Aplet et al. 1998), ecophysiology (Vitousek et al. 1990, Cordell et al. 1998), weathering versus atmospheric sources of nutrients (Cochrane and Berner 1997, Vitousek et al. 1999), biological N fixation (Kurina and Vitousek 1999, 2001; Crews et al. 2001), nutrient limitation (Raich et al. 1996), and most recently as a testbed for the widely used Century ecosystem model (Raich et al. 2000).

A Substrate Age Gradient across the Hawaiian Islands

For the rest of this book, I focus on a substrate age gradient across the Hawaiian Islands. I emphasize this gradient for several reasons. First, the process of ecosystem development is inherently an interesting one—it is no coincidence that "primary succession," the development of a biological community on previously unoccupied land, was the subject of much of the earliest recognizable terrestrial ecology (Cowles 1899, Cooper 1923). Second, large differences in the sources, losses, cycling, and availability of nutrients can be expected as sites develop from raw rock to ancient, deeply leached soils—and these patterns should yield insight into the mechanisms that control nutrient cycling and limitation. Indeed, there is a well-developed theory describing causes of variation in soil nutrient pools and nutrient availability during ecosystem development (Walker and Syers 1976), and that theory has been extended to explain the geographical distribution of nutrient limitation (Vitousek and Sanford 1986). Third, properties of these particular systems lend themselves to tracing the sources of nutrients, as discussed below. Finally, we know a good deal about these sites; the youngest and oldest in particular have supported long-term experiments that evaluate the nature and consequences of nutrient limitation.

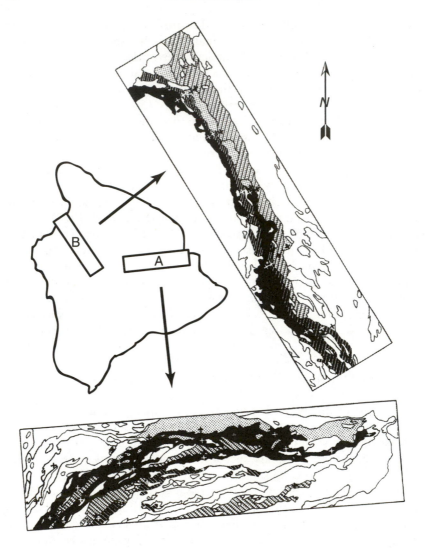

FIGURE 3.6. The Mauna Loa environmental matrix, from Vitousek et al. (1992). A young flow (< 140 yr, in black) versus an old flow (> 2.8 ky, stippled) of 'a'ā (diagonal lines) versus pāhoehoe textures were sampled across a wide range of elevations on the wet east flank (rectangle A) versus the dry northwest flank (rectangle B) of Mauna Loa volcano, yielding a complete matrix of age, elevation (and thus temperature), precipitation, and lava flow texture.

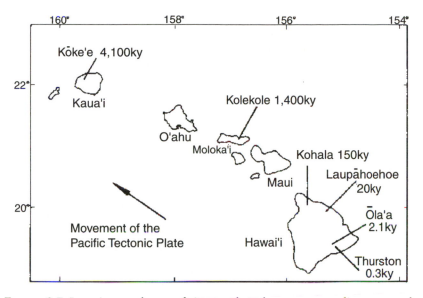

FIGURE 3.7. Locations and ages of sites on the substrate age gradient across the Hawaiian Islands (Crews et al. 1995); this gradient is the focus of most of the rest of this book.

This substrate age gradient makes use of the increasing age of Hawaiian volcanoes with increasing distance from the hot-spot, from southeast to northwest across the islands (fig. 2.1). Originally, much of the gradient was developed by Ralph Riley for a study of the regulation of nitrogen trace gas emissions (Riley and Vitousek 1995, Riley 1996); it was later refined by Tim Crews, Darrell Herbert, and Kanehiro Kitayama to consist of six sites that range in age from ~300 to ~4.1 million years. These include the 0.3 ky (ky = 1000 years) Thurston site on Kīlauea volcano, the 2.1 ky ʻŌlaʻa site on Mauna Loa, the 20 ky Laupāhoehoe site on Mauna Kea, the 150 ky Kohala site on Kohala Volcano, the 1400 ky Kolekole site on East Molokaʻi volcano, and the 4100 ky Kōkeʻe site on Kauaʻi (fig. 3.7) (Crews et al. 1995). Not every site has been included in each study along the gradient, and additional sites have been included for particular studies.

All of the core sites are between 1120 and 1210 m elevation, and so experience a mean annual temperature near 15.5°C, and all have mean annual precipitation close to 2500 mm (table 3.1). This combination of elevation and precipitation was selected because lower and drier areas have been altered far more extensively by human activity, in many cases initially by the Polynesian discoverers of Hawaiʻi. In contrast, none of the core sites has been cleared or (as far as we can tell) systematically altered

TABLE 3.1
Characteristics of sites on the substrate age gradient across the Hawaiian
Islands. In addition to the information here, all sites have mean annual
temperatures of 15.5–16° C, and all average ~2500 mm/yr of precipitation.
Any readers who have been struggling with Hawaiian words should note
the terminology for soils, and be grateful.

Site	Substrate age (ky)	Elevation (m)	Island	Volcano	Soil classification
Thurston	0.3	1176	Hawai'i	Kīlauea	Lithic Hapludand (Andisol)
'Ōla'a	2.1	1200	Hawai'i	Mauna Loa	Thaptic Udivitrand (Andisol)
Laupāhoehoe	20	1170	Hawai'i	Mauna Kea	Hydric Hapludand (Andisol)
Kohala	150	1122	Hawai'i	Kohala	Hydric Hydrudand (Andisol)
Kolekole	1400	1210	Moloka'i	East Moloka'i	Hydric Hydrudand (Andisol)
Kōke'e	4100	1134	Kaua'i	Kaua'i	Plinthic Kandiudox (Oxisol)

by direct human action. Wetter sites tend towards bogs on the older substrates, including those on a parallel substrate age gradient across the
Islands established by Kanehiro Kitayama (Kitayama and Mueller-Dombois
1995, Kitayama et al. 1997).

All of the age gradient sites are located on the constructional surface of
shield volcanoes, where the influence of erosion has been minimal; on
older mountains, we selected sites on constructional surfaces that remain
in inter-stream areas. *Metrosideros polymorpha* is the dominant tree on
every site, and many other species occur across the gradient. Finally, the
underlying rock in all of the sites is basalt, as it is for almost all of the
Hawaiian Islands.

Overall, the climate, organism, parent material, and relief state factors
(sensu Jenny 1980) are similar across these sites, while time (substrate age)
ranges over four orders of magnitude, from 0.3 to 4100 ky. In terms of our
ability to control other sources of variation, I believe that these six sites
represent the best-defined long age sequence of sites on Earth. Nevertheless, it is far from perfect; there are assumptions, uncertainties, and simpli-
fications that go into any study based on environmental gradients. I will
describe two of those here—how we determine and define the substrate

age for a site, and the influence of systematic variations in climate history among sites. A more complete analysis of between-site variations and uncertainties in the major factors controlling ecosystem properties and processes across this gradient is accessible at www.stanford.edu/vitousek/princetonbook.html.

Age Control

The substrate age gradient is intended to represent a gradient in time—and thus knowing the time at which surface biological and geochemical processes began to influence each site is crucial. For the stable constructional surface of a shield volcano, that time should be the age of the substrate (lava flow or deep ash deposit) that underlies the site. That sounds straightforward, but it is not always so. For example, we know the history of the Thurston site quite well; it is on the flank of a satellite volcanic shield of Kīlauea Volcano that erupted actively between about 1450 and 1500 AD (Clague et al. 1999). The lava is tens of meters thick; no vegetation or soil organic matter now on the site predates that eruption. In the first 250–300 years following the eruption, vegetation colonized the site and a thin soil developed, made up of organic matter and volcanic tephra from the nearby vents of Kīlauea. Finally, an explosive eruption in 1790 killed the forest (along with a Hawaiian army that was passing through the area), deposited a thicker layer of tephra, and re-initiated vegetation succession—this time with a legacy of some buried soil organic matter. The site is young, only hundreds of years old; we call it 300 years old (0.3 ky) because it must be between 200 and 500 years old, and we have to call it something.

The next-youngest site, 'Ōla'a, is covered with a thicker blanket of the same 1790 tephra that covers Thurston. However, this 1790 tephra overlays deep layers of older tephra, the oldest and thickest of which is ~2100 yrs old. Trees at 'Ōla'a are much larger than at Thurston; we believe that many survived the 1790 tephra, and in any case their roots reach into the deep tephra layers. We define this site to be 2.1 ky old, although certainly many of our soil measurements are shaped by the 1790 tephra layer. The older sites no doubt had similarly complicated histories; we know that is true of the 20 ky Laupāhoehoe site (Kennedy et al. 1998, www.stanford.edu/vitousek/princetonbook.html). However, the older the site, the less that uncertainties of hundreds or even thousands of years are relevant.

Climate History

I suspect that the most important—and most interesting—source of variation among sites (other than substrate age) is differences in climate history. The chronosequence approach not only assumes that sites have the same current climate; it further assumes that older sites had the same climate (and other conditions) as the younger ones, when they were young themselves. For example, it assumes that when our 4100 ky site was 0.3 ky old, it had the same climate that the 0.3 ky one does now. That is not true here; I doubt that it is true anywhere on Earth.

Hotchkiss et al. (2000) evaluated past variations in the climate of the age-gradient sites, concentrating on three sources of change over time:

1. Differences due to the local influence of global climate change. Although the Hawaiian Islands' maritime tropical location buffers them against the most extreme variations in Earth's glacial-interglacial cycle, palynological and geomorphological analyses show that montane forest areas of the islands were cooler and substantially drier during glacial times, and the trade wind inversion was at a lower altitude (Porter 1979, Gavenda 1992, Hotchkiss 1998, Hostetler and Clark 2000, Hotchkiss in press). While the youngest sites have experienced only relatively warm, wet interglacial conditions like the present, the older sites have been through multiple glacial-interglacial cycles.

2. Differences due to island subsidence. As described in chapter 2, the mass of a growing volcano depresses the underlying oceanic crust, leading to subsidence. Consequently, the final elevation of a site—once subsidence has ceased, a million years or more after a volcano is built—can be 1200–2000 meters lower than its initial elevation (Ludwig et al. 1991, Moore and Clague 1992, Price 2002). Much of the early history of the older sites was spent at a higher, cooler, drier elevation. Moreover, if the youngest sites were to persist without being covered by additional lava for another million years, they would no longer be montane sites; they might even be submerged.

3. Differences due to upwind erosion, which can expose sites that were in leeward rain shadow positions to much greater precipitation. For example, the 4100 ky Kōke'e site began as a drier leeward site, but the removal of much of the island's windward side by sliding and fluvial erosion has exposed it to the trade winds and caused increased precipitation there (Hotchkiss et al. 2000).

The combined influences of these processes on the climate of the 1400 ky Kolekole site are summarized in fig. 3.8 (Hotchkiss et al. 2000). This site was substantially cooler and drier through most of its history than it is at present—or than the younger sites are now. The 4100 ky Kōke'e site probably has experienced even more dramatic changes—but the first million and a half years of its history were in the Pliocene, of which we know next to nothing (in terms of Hawaiian climates).

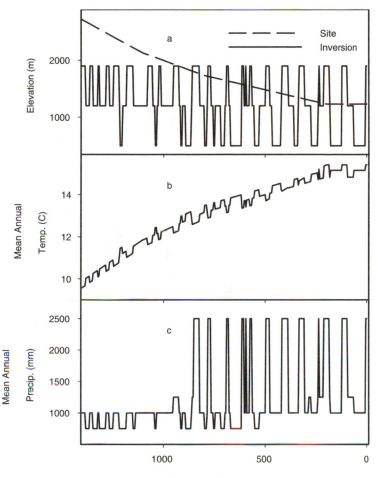

FIGURE 3.8. A proposed climate history for the 1400 ky Kolekole site on the substrate age gradient, modified from Hotchkiss et al. (2000). (a) Changes in the elevation of the site due to subsidence (dashed line), and in the elevation of the tradewind inversion resulting from glacial-interglacial climate cycles (solid line). (b) Changes in mean annual temperature. (c) Changes in mean annual precipitation; rainfall is much greater below the tradewind inversion. In each case, climate is assumed to oscillate between warm wet interglacial, intermediate interstadial, and cold dry glacial conditions. This analysis suggests that for much of its history, the site has been substantially cooler and drier than at present.

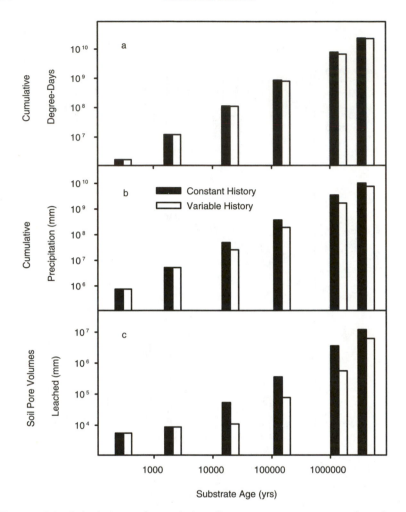

FIGURE 3.9. Calculations of cumulative climate parameters across the substrate age gradient, assuming constant climate (solid bars), or taking into account subsidence, glacial-interglacial climate change, and (in the oldest site) erosion of the upwind portion of the island (open bars). Modified from Hotchkiss et al. (2000). (a) Cumulative degree days over the history of each site. (b) Cumulative precipitation in mm. (c) Cumulative leaching of water through the soil of each site (precipitation minus evapotranspiration). Accounting for climate history greatly reduces calculated leaching in particular (note the log scale of the y-axis) relative to assuming a constant climate, but does not change the monotonic nature of the sequence.

Hotchkiss et al. (2000) also calculated climatic histories of sites along the age gradient, using cumulative degree-days accumulated through stand development (cf. Johnson et al. 2000) for temperature and the cumulative volume of precipitation inputs and of water leaching through the soil. In this analysis, leaching in particular differs substantially from what would be expected given constant climate—but the sites remain a monotonic sequence (fig. 3.9), and the differences among them are far from subtle. Accordingly, I will treat this set of sites as a simple gradient in substrate age, and use it as such in analyses of the regulation of nutrient availability, cycling, and limitation.

Basic Features of the Gradient

SOILS

Soils in the five younger sites are classified as Andisols, but the 4100 ky site is an Oxisol (table 3.1). Andisols are influenced by volcanic parent material and its weathering products; Oxisols are highly weathered soils that are widespread in the humid tropics (Sanchez 1976). Among the Andisols, the two youngest have poorly developed profiles and tend towards Inceptisols; the 20 ky and 150 ky sites are classic Andisols dominated by gel-like noncrystalline secondary minerals; and the 1400 ky site tends towards a highly weathered Ultisol. The most abundant soil minerals change substantially along the sequence (Vitousek et al. 1997b). Soils of the two youngest sites are dominated by primary minerals (olivine, glass, and plagioclase feldspar) derived from the volcanic parent material, but these minerals weather rapidly, nearly disappearing by the 20 ky site (fig. 3.10). The young sites also contain the secondary non-crystalline minerals ferrihydrite, allophane, and imogolite; these form as weathering products of primary minerals (Shoji et al. 1993), and become the dominant minerals in the 20 ky and 150 ky sites (Chorover et al. 1999). Allophane, imogolite, and ferrihydrite are metastable, X-ray amorphous minerals with large, reactive, variable-change surfaces (cf., Uehara and Gillman 1981) that strongly bind phosphorus and soil organic carbon (Wada 1989, Schwertmann and Taylor 1989). Over long time scales, these minerals continue to weather, eventually forming the secondary kaolin and crystalline sesquioxide minerals that characterize highly weathered tropical soils. These crystalline minerals are less reactive; they become important in the 1400 ky site and dominate the clay fraction of the oldest site on the sequence (fig. 3.10).

VEGETATION

The structure and composition of vegetation along the age gradient were determined by Kanehiro Kitayama (Crews et al. 1995, Kitayama and Mueller-Dombois 1995). The height of the upper canopy increases from

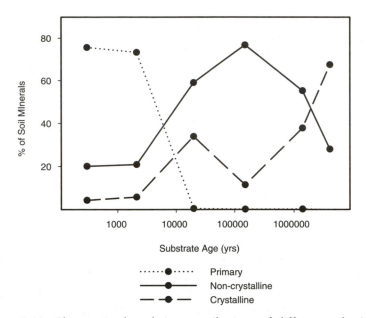

FIGURE 3.10. Changes in the relative contributions of different soil minerals along the substrate age gradient, revised from Vitousek et al. (1997b). Primary minerals (the original minerals that were inherited from parent material) dominate the youngest sites, followed by an assemblage of pedogenic non-crystalline and poorly-crystalline minerals (allophane, imogolite, ferrihydrite) in intermediate-aged sites; these in turn are replaced by crystalline minerals and hydroxides (kaolinite, gibbsite) in the oldest sites.

the youngest into intermediate-aged sites, and then declines substantially in the older sites (fig. 3.11). *Metrosideros polymorpha* overwhelmingly dominates all sites, accounting for > 75 percent of tree cover, and several additional species occur in most or all sites. Native tree ferns in the genus *Cibotium* (*C. glaucum* and *C. chammissois*) dominate the subcanopy in the younger sites, but decline monotonically to very low abundance in the oldest site (fig. 3.12). Plant species diversity is least in the youngest site (28 species in 0.2 ha) and greatest in the oldest (66 species/ in 0.2 ha); intermediate sites display no consistent pattern (Crews et al. 1995). In subsequent chapters, I will discuss how these changes in soils and vegetation relate to changes in nutrient cycling and limitation across the age gradient.

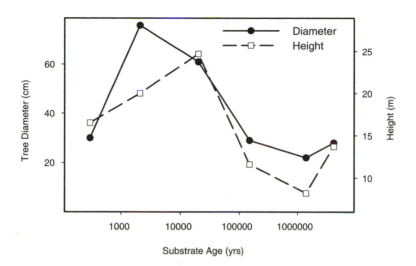

FIGURE 3.11. Changes in the stature of dominant *Metrosideros polymorpha* trees along the substrate age gradient, modified from Kitayama and Mueller-Dombois (1995). Heights and diameters represent the means of the five largest trees in each plot.

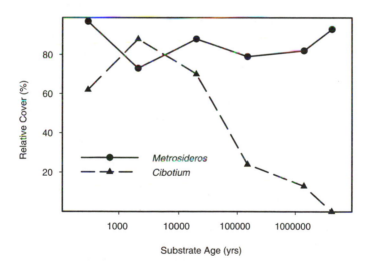

FIGURE 3.12. The relative dominance of tree cover by *Metrosideros polymorpha* and of understory cover by *Cibotium* tree ferns along the Hawaiian age gradient, from information in Crews et al. (1995). *Metrosideros* accounts for > 75% of the tree cover in all six sites.

Chapter Four

PATTERNS AND PROCESSES IN

LONG-TERM ECOSYSTEM DEVELOPMENT

A Theory for Nutrient Dynamics during Ecosystem Development

How would we expect nutrient availability, cycling, and limitation to vary as ecosystems develop over millions of years, from new rock to very old substrates? How would organisms be expected to respond to any variations? Based on their studies of soil properties on several developmental sequences in New Zealand, the soil scientists T. W. Walker and J. K. Syers (1976) suggested that because P is present in rocks and lacks a significant gas phase, ecosystems developing on new substrates (following dune stabilization, volcanic eruption, etc.) contain all the P that they will ever have, mostly as the primary mineral apatite. When apatite is exposed to water and acidity, it weathers and releases P as biologically available phosphate (PO_4). Some of this PO_4 is taken up by plants and animals, returns to soil as organic P, and can be mineralized back to PO_4; some is precipitated or adsorbed by secondary minerals within the soil.

Due both to biological demand for P and to the very slow reversibility of P adsorption and precipitation, P moves very slowly through soils (Wood et al. 1984). However, a small fraction of soil P can be lost from ecosystems as dissolved inorganic or organic P; another small fraction can be precipitated in wholly insoluble and/or physically protected forms (e.g., the interior of stable soil aggregates), and effectively "lost" to the biota of a system in this way. Over time, therefore, the quantity of P in the system as a whole should decline, and an increasing fraction of the remaining P should be bound up in the insoluble/physically protected fraction. Ultimately, an ecosystem could reach what Walker and Syers (1976) called a "terminal steady state" in which little P is present in forms that can be cycled through organisms, and P limitation to biological processes is profound. Walker and Syers (1976) summarized these dynamics of soil P graphically (fig. 4.1); the New Zealand chronosequences they studied fit the pattern in fig. 4.1 very well.

Walker and Syers (1976) further pointed out that in contrast to P, N is absent from most primary substrates. Accordingly, biologically available

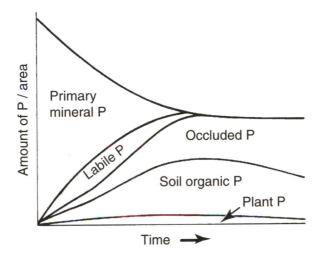

FIGURE 4.1. The Walker and Syers conceptual model for changes in P fractions during long-term soil development, revised from Walker and Syers (1976). The proportion of parent material P that remains (upper line) decreases over time due to leaching losses of P. The dominant fractions of soil P vary from primary mineral P in young soils; to labile P and organic P in intermediate-aged soils; to insoluble (occluded) P and recalcitrant organic P in old soils.

N should be in relatively short supply early in soil development—and organisms with the capacity to fix N_2 from the vast pool in the atmosphere should have a substantial advantage over those that cannot fix N. Indeed, many (although not all) primary successions include an early stage dominated by one or more symbiotic N fixers (Crocker and Major 1955, Van Cleve et al. 1991, Chapin et al. 1994, Schlesinger et al. 1998). Where N fixers are absent, N must be accumulated from fixed N in precipitation, and rates of N supply by this pathway are low, especially in regions remote from human influence (Galloway et al. 1995). Consequently, N supply would be expected to limit biological processes early in ecosystem development. Walker and Syers (1976) suggested that N accumulates (rapidly from biological fixation or more slowly from precipitation) until it equilibrates with the supply of P, at a stoichiometric ratio equivalent to organisms' requirements for N and P (Reiners 1986, Sterner and Elser 2002). At and above that point, there is little or no advantage to acquiring N by fixation or retaining it within ecosystems. This N:P ratio often is near 15:1 by mass, although it can vary among organisms (Elser et al. 1996).

Overall, Walker and Syers (1976) suggested that plant growth in ecosystems on young soils might be expected to be N-limited, while growth on old soils should be P-limited; in intermediate-aged sites N and P should equilibrate at higher levels of supply. This developmental perspective leads to a geographical one. Through the Pleistocene, soil development in temperate and boreal regions has been reset frequently by glaciation and periglacial processes, while the soils of geomorphically stable tropical landscapes have remained more or less intact. Consequently, Walker and Syers' model suggests that the productivity of many tropical systems should be limited by P supply, but that of most temperate and boreal systems should be limited by N. Direct measurements show that the latter expectation generally is met (Lee et al. 1983, Tamm 1991, Vitousek and Howarth 1991); less direct observations based on element ratios in plants and soils generally support the former (Vitousek and Sanford 1986).

A number of subsequent papers have tested, extended, and modified Walker and Syers' model. J. Walker et al. (1983) demonstrated progressive translocation of P to deep B-horizons that lie below the reach of plant roots along a sand dune sequence in Australia. Smeck (1985) and Cross and Schlesinger (1995) showed that more intensively leached soils (e.g., Oxisols) retain little primary mineral P and have more P in insoluble and physically protected forms in comparison to less-leached soil orders. Lajtha and Schlesinger (1988) demonstrated that P fractions changed relatively little along an arid chronosequence in New Mexico, and suggested that the Walker and Syers (1976) model applies primarily to soils in mesic to wet climates. Chapin et al. (1994) and Schlesinger et al. (1998) found evidence for P limitation on young soils dominated by symbiotic N fixers, and suggested that where N fixers are dominant, weathering rates might be too slow to supply enough P. Finally, Dahlgren (1994) and Holloway et al. (1998) demonstrated that many sedimentary rocks contain fixed N that can be supplied to soils and ecosystems, even early in soil development, and Newman (1995) showed that the importance of atmospheric inputs of P to ecosystems should not be neglected.

While many of these studies contribute extensions, constraints, and mechanisms to Walker and Syers (1976), they do not diminish the value of the underlying conceptual model. However, few studies have tested the implications of the model for nutrient limitation or other biological processes—and none have evaluated interactions between biotic and abiotic processes across a full developmental sequence. The Hawaiian age gradient offers the opportunity to (1) evaluate patterns in soil P fractions during long-term soil development; (2) understand the implications of these patterns for the availability of P and other nutrients; (3) explore how productivity and other biological processes change during ecosystem development; (4) analyze feedbacks between plants, microorganisms, and the geochemical processes controlling nutrient availability, as they vary from young to old soils; and (5) determine the nature and consequences of nutrient limitation during long-term ecosystem development. I address the first four of these issues as they apply to the Hawaiian age gradient in this chapter, and describe experimental studies of nutrient limitation in the next chapter.

BIOGEOCHEMICAL PROCESSES ON THE SUBSTRATE AGE GRADIENT

Soil P Pools

Crews et al. (1995) determined the forms of soil P in all the sites on the gradient using the Hedley sequential extraction procedure (Hedley et al. 1982, Tiessen and Mohr 1993). This procedure yielded eleven fractions, which were combined into four pools, termed primary mineral, organic, labile, and occluded, to approximate the pools described by Walker and Syers (1976) (fig. 4.1). Losses of P from the soil profile were calculated with reference to an immobile index element, as described in chapter 6 (Chadwick et al. 1990).

Substantial quantities of P were lost from the soils over time (fig. 4.2a) (Vitousek et al. 1997b). However, over most of soil development a roughly equivalent fraction of total soil mass was lost as well, so that the quantity of P in the upper 50 cm remained roughly constant. Primary mineral P (apatite) accounted for most of the soil P in the youngest site, but virtually disappeared by 20 ky; insoluble and/or protected P (occluded P) constituted the majority of remaining soil P in the two oldest sites (fig. 4.2b). Non-occluded P—forms that could become available to organisms on some reasonable time scale—decreased in the older sites, but not as dramatically as the model (fig. 4.1) suggests. Given the inherent ambiguity of comparing operationally defined P pools from the Hedley fractionation with the conceptually defined pools of the Walker and Syers model, the model does a reasonably good job of matching observed changes in soil P pools across the sites.

C and N Pools

Organic C and N pools in soils accumulated rapidly early in soil development, with the youngest site accumulating on average ~500 and 30 kg ha^{-1} yr^{-1} of C and N, respectively, during its ~300 yr history. Soil C and N pools continued to increase to the intermediate-aged sites, and then declined into the oldest site (fig. 4.3) (Crews et al. 1995, Torn et al. 1997). The increase from young to intermediate-aged sites reflects both continued accumulation in these developing systems and the stabilization of organic matter by adsorption to the highly reactive non-crystalline minerals that dominate intermediate-aged sites (fig. 3.10); the decline from intermediate to old sites reflects the weathering of non-crystalline minerals to form

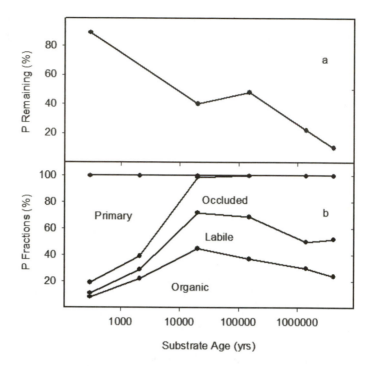

FIGURE 4.2. Losses of P and changes in soil P fractions across the Hawaiian substrate age gradient, based on Crews et al. (1995) and Vitousek et al. (1997). (a) The fraction of original parent material P remaining in each site, calculated with reference to an immobile index element as described in chapter 6. (b) Changes in the proportional distribution of P among chemical fractions. Loss of parent material P continues across the gradient, and an increasing fraction of the remaining P is in occluded forms.

less-reactive kaolin and sesquioxide (Uehara and Gillman 1981, Torn et al. 1997).

Available Nutrients

Measuring nutrient availability in soils is not straightforward. Inorganic forms of nutrients (e.g., NH_4^+, NO_3^-, PO_2^{3-}, Ca^{2+}) that are in solution or exchangeably bound to soil colloids or organic matter can be extracted and measured, but chemical extractants do not duplicate the ability of plants and microbes to access soil nutrients. Moreover, some dissolved organic forms of nutrients (N in particular) can be taken up by plants as well as microbes (Chapin et al. 1993, Näsholm et al. 1998). More im-

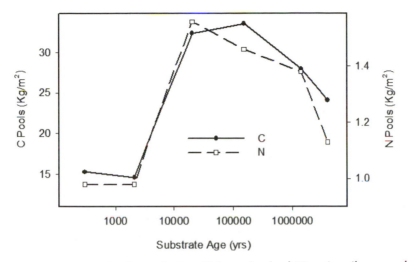

FIGURE 4.3. The pools of organic C and N to a depth of 50 cm in soils across the Hawaiian age gradient, from information in Crews et al. (1995).

portantly, where demand for a nutrient is high, pools of available nutrients in the soil turn over extremely rapidly, often in less than a day (Stark and Hart 1997, Olander and Vitousek in press). In such cases, the rate of nutrient supply (through mineralization, solubilization, and desorption) into the biologically available pool provides a more useful measure of available nutrients than does the pool size at any one time. However, rates of supply are difficult to measure without affecting the processes involved.

We used a number of techniques to estimate available nutrients in Hawaiian soils. Exchangeable cations (Ca, Mg, K) were determined using a standard ammonium acetate extract. For P, we used anion exchange resins in the field, in addition to the Hedley fractionations described above. Finally, for N we used anion and cation exchange resins in the field, extractions of dissolved and exchangeable pools of NH_4^+ and NO_3^-, and measurements of net and gross rates of N mineralization and nitrification.

Exchangeable cation concentrations vary straightforwardly along the sequence; all are relatively large in the two youngest sites, then decline (fig. 4.4a), although K increases again in the oldest site. Resin-extractable P is low in the youngest site, where most soil P resides in primary minerals; it increases in the intermediate-aged sites (peaking in the 150 ky Kohala site), and then declines in the older sites (fig. 4.4b), where most remaining P is in insoluble and/or physically protected forms (Vitousek et al. 1997b). By any of our measures, the two youngest sites have lower N availability than the four older sites; differences among the older sites are

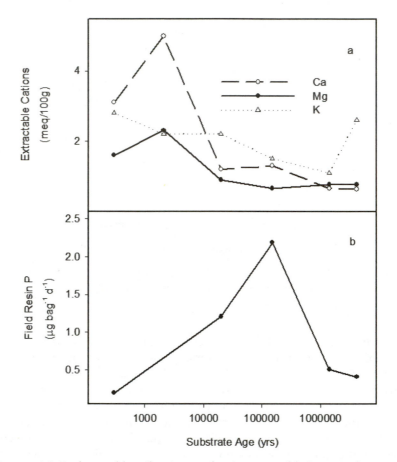

FIGURE 4.4. Exchangeable soil cations and resin-extractable P across the age gradient, modified from Crews et al. (1995) and Vitousek et al. (1997b). (a) Exchangeable cations generally decline from the young to the old sites. (b) Extractable P concentrations from resin bags incubated in the field peak in the intermediate-aged sites.

inconsistent across methods (Riley and Vitousek 1995, Crews et al. 1995, Hedin et al. 2003) (fig. 4.5). Overall, young sites have high levels of available cations and low available N and P; intermediate-aged sites have low cations and relatively high N and P; and the oldest sites have high N but low cations and P.

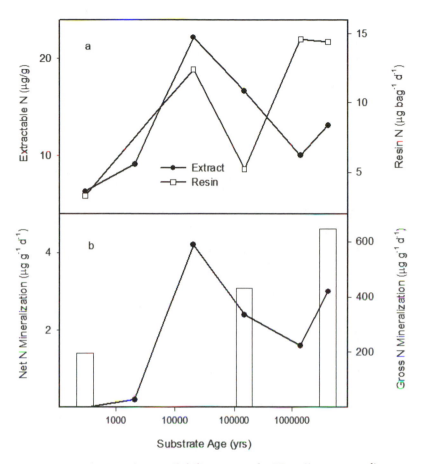

FIGURE 4.5. Measures of N availability across the Hawaiian age gradient, synthesized from Riley and Vitousek (1995), Crews et al. (1995), and Hedin et al. (2003). (a) KCl-extractable and resin-extractable ammonium-N plus nitrate-N. (b) Net N mineralization in laboratory incubations (line) and gross N mineralization measured by ^{15}N isotope dilution (bars). Although there is substantial variation among the various measures of N availability, all yield the lowest N pools and rates of transformations in the two youngest sites.

Foliar Nutrients

An alternative way to assess nutrient availability across a range of sites is to ask the plants. Foliar nutrient concentrations have long been used to indicate plant nutrient status, and even the likelihood of limitation by particular nutrients (Van den Driessche 1974, Medina and Cuevas 1994,

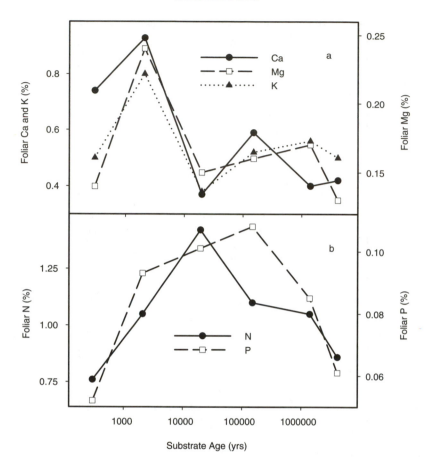

FIGURE 4.6. Foliar nutrient concentrations in sun leaves of *Metrosideros poly-morpha* along the Hawaiian substrate age gradient, modified from Vitousek et al. (1995b). (a) Cation concentrations; Ca in particular decreases from young to old sites. (b) N and P concentrations peak in the intermediate-aged sites.

Bruijnzeel and Veneklass 1998). Factors other than soil nutrient availability can cause variation in foliar nutrient concentrations (Chapin et al. 1986, Reich et al. 1992, Tanner et al. 1992). However, where leaves of the same age and canopy position from a single species are sampled on a range of sites that differ substantially in nutrient availability (as here), foliar nutrients reflect soil nutrient availability reasonably well. We collected *Metrosideros* foliage from upper-canopy, full-sun positions across all the sites, shooting down small branches with a shotgun in the taller canopies or with a slingshot in the shorter ones. Several collections were made in each site at different times of the year, and nutrient concentrations were consistent across all collections.

Foliar nutrient concentrations in *Metrosideros* vary markedly across the sites, generally in synchrony with available nutrients in soil. Ca concentrations are relatively high in the two youngest sites, then decline into the older ones; Mg and K are less consistent, except for high concentrations in the 2.1 ky site (fig. 4.6a). Foliar P is low in the young sites, nearly two-fold greater in intermediate-aged sites, and low again in the oldest site (fig. 4.6b)—just like resin-extractable P in soil (fig. 4.4b). Foliar N also is low in the youngest sites and substantially higher in the intermediate-aged sites (fig. 4.6b), reflecting soil N availability (fig. 4.5). However, foliar N concentrations drop again in the oldest site—unlike available soil N, which remains high. Foliar N might be lower in the old site because P availability is lower, and *Metrosideros* maintains a fairly consistent stoichiometric ratio of N and P (10.1–14.5 by mass) in its leaves across all the sites. Similar patterns of variation in foliar nutrients occur in several other species that are found in all of the sites (Vitousek et al. 1995b).

Forest Productivity

Herbert and Fownes (1999) determined aboveground net primary production (ANPP) in five sites (all except the 2.1 ky 'Ōla'a site) by measuring aboveground litterfall (leaves and twigs) and stemwood increment over an annual cycle. ANPP ranged from nearly 1400 g m^{-2} yr^{-1} in the 1400 ky site to < 800 g m^{-2} yr^{-1} in the 4100 ky site (table 4.1), with no strong relationship to nutrient availability. However, leaf litterfall (annual leaf fall) tracked N and/or P availability across the sites, being lower in the youngest and oldest sites than in the more-fertile intermediate-aged sites (fig. 4.7). Herbert and Fownes (1999) also measured soil CO_2 evolution through the year. Assuming that the pool of C in litter and soil is in steady state, outputs from the soil (in CO_2 evolution, plus small dissolved organic C outputs) must be balanced by inputs (litterfall from aboveground production, root turnover, and root/mycorrhizal respiration). Given these assumptions, soil CO_2 evolution should thus be a measure of ANPP plus total belowground C allocation (Raich and Nadelhoffer 1989)—and like leaf litterfall, CO_2 emission peaked in intermediate-aged sites on the sequence (fig. 4.7).

Herbert and Fownes (1999) subtracted above-ground litterfall C from CO_2–C emissions to obtain total belowground C allocation (Raich and Nadelhoffer 1989); they then estimated belowground net primary production (BNPP) by assuming that half of belowground C allocation is root/mycorrhizal respiration, and the remainder BNPP. They found values

TABLE 4.1

Components of productivity in sites along the Hawaiian age gradient. Calculated belowground production was determined by mass balance according to Raich and Nadelhoffer (1989), assuming that half of belowground C allocation goes to root respiration and the rest to belowground production. Most values from Herbert and Fownes (1999), except root production, which is from Ostertag (2001). Values in g m^{-2} yr^{-1}

	0.3 ky	20 ky	150 ky	1400 ky	4100 ky
Leaf litterfall	382	430	506	513	308
Twig litterfall	166	118	133	360	116
Wood increment	506	422	270	522	322
Total aboveground	1054	970	909	1395	746
Calculated belowground	524	634	873	568	683
Root production	173	169	—	—	75
Total productivity	1578	1604	1782	1963	1429
CO_2–C evolution	798	908	1192	1004	896

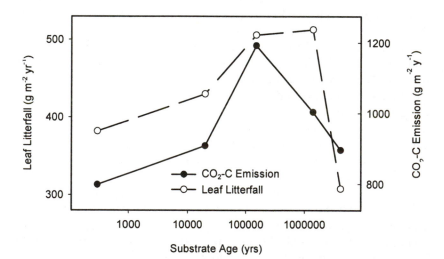

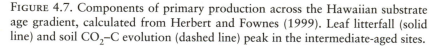

FIGURE 4.7. Components of primary production across the Hawaiian substrate age gradient, calculated from Herbert and Fownes (1999). Leaf litterfall (solid line) and soil CO_2–C evolution (dashed line) peak in the intermediate-aged sites.

ranging from 524–873 g m^{-2} yr^{-1} (table 4-1). Ostertag (2001) estimated BNPP in three of the sites using a different method. First, she repeatedly measured the biomass of live and dead roots through an annual cycle—finding no significant differences within sites among sampling dates. She then measured rates of root decomposition, and calculated the root production necessary to maintain the observed stocks of roots in the face of the observed rates of decomposition. BNPP by this method ranged from 75 to 173 g m^{-2} yr^{-1}, with the 4100 ky site having significantly lower production than the younger sites (table 4.1) (Ostertag 2001). The differences between these two approaches are substantial, and the reasons for them are unclear. BNPP is notoriously difficult to measure under the best of circumstances; in this case, both approaches require different steady-state assumptions that are difficult to test.

Overall, there is no clear pattern of variation with substrate age for many of the components of NPP across the soil age gradient (table 4.1)—and where there is interpretable variation (as in leaf litterfall and in CO_2–C emissions), the lack of within-age replication precludes any strong tests of statistical significance. In contrast, we observe substantial differences in primary productivity across climatic gradients in Hawai'i (Raich et al. 1997, Schuur and Matson 2001, Austin 2002).

Efficiencies of Resource Use

How can forest productivity remain relatively constant across the gradient, in the face of much greater changes in nutrient availability? Three different mechanisms could be important:

1. Even the low levels of nutrients in infertile young and old sites could be sufficient to maintain forest productivity; nutrient supply could be non-limiting in all of the sites.
2. Forests could maintain productivity by acquiring nutrients more effectively in low nutrient sites, through greater allocation of resources to roots, mycorrhizal symbionts, extracellular enzymes, or other pathways.
3. Forest could use nutrients more efficiently in low nutrient sites, producing more biomass per unit of nutrient.

The experimental studies discussed in chapter 5 unequivocally show that nutrient supply limits NPP in the youngest and oldest sites, ruling out the first possibility. For the second, Kathleen Treseder found trends towards greater investment in nutrient acquisition in the infertile youngest and oldest sites (Treseder and Vitousek 2001a). The low-P 4100 ky site supported more mycorrhizal roots that had a greater capacity to transport P and greater production of extracellular phosphatases than did

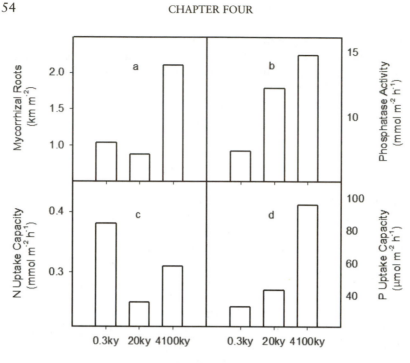

FIGURE 4.8. Investment in nutrient acquisition in three sites along the Hawaiian age gradient, from information in Treseder and Vitousek (2001a). (a) Mycorrhizal root length. (b) Root-associated phosphatase enzyme activity. (c) N uptake capacity. (d) P uptake capacity. Investment in P acquisition is greatest in the old site, and the one index of investment in N acquisition is greatest in the young site.

other sites (fig. 4.8a, b, d); roots in the low-N 0.3 ky site had a greater capacity to transport N than roots in other sites (fig. 4.8c). However, although plants in low-nutrient sites might work harder to get their nutrients, they still don't acquire nearly as much as do plants in the more-fertile sites. Both N and P concentrations in leaves (fig. 4.6b) and the quantities of these nutrients that cycle between plants and soils annually (fig. 4.9) are substantially less in the low-nutrient young and old sites than in the more fertile intermediate-aged sites (Vitousek et al. 1995b, Herbert and Fownes 1999). A greater investment in nutrient acquisition might keep forests on infertile soils going, but the most important reason that forests on such sites are relatively productive is that they use the nutrients that they do acquire more efficiently.

Nutrient-use efficiency is a much-discussed concept, and a number of different definitions and indices of nutrient-use efficiency have been suggested

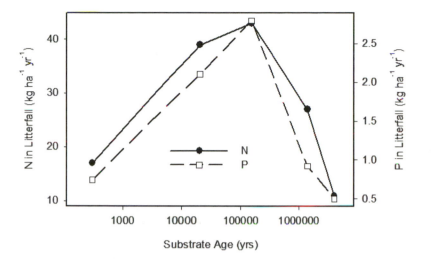

FIGURE 4.9. The quantity of N and P circulating annually through leaf litterfall in sites across the substrate age gradient, from Herbert and Fownes (1999).

(Chapin 1980, Vitousek 1982, Berendse and Aerts 1987, Bridgham et al. 1995, Aerts and Chapin 2000, Hiremath and Ewel 2001). At the level of forest stands, the most appropriate of these are production per unit of nutrient taken up (Berendse and Aerts 1987) and production per unit of nutrient available in soil (Bridgham et al. 1995). While these differ conceptually in some important ways, both yield qualitatively similar patterns for the Hawaiian age gradient sites, because N and P uptake generally correlate strongly with N and P availability (figs. 4.4b, 4.5, 4.9). I discuss nutrient-use efficiency in terms of production per unit of nutrient taken up here; this definition allows me to focus on the mechanism controlling nutrient-use efficiency within trees.

How can a tree (and by extension a forest) use nutrients more efficiently than another? Berendse and Aerts (1987) suggested that two components could be involved: (1) nutrient productivity, which is production per unit of nutrient within plants at a given time; and (2) the mean residence time of a unit of a nutrient within plants. I discuss nutrient productivity in terms of NPP per unit of nutrient (N or P) in the forest canopy (Ågren and Bosatta 1996); the canopy is the locus of energy capture, and so the place where mechanisms controlling nutrient productivity should be expressed.

N and P productivity both increased approximately twofold across the substrate age gradient (fig. 4.10). Greater nutrient productivity contributed substantially to overall nutrient-use efficiency in the low-P 4100 ky site, which had the smallest leaf area and thus the least within-canopy shading

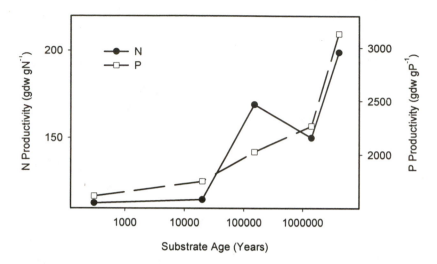

FIGURE 4.10. N and P productivity increase across the Hawaiian age gradient, modified from Herbert and Fownes (1999). Nutrient productivity is calculated as NPP (in grams dry weight, or gdw) per unit of each nutrient in the forest canopy.

of these sites (Herbert and Fownes 1999), but it made no positive contribution to nutrient-use efficiency in the young site. In contrast, a longer residence time of nutrients within the canopy contributed to the nutrient-use efficiency of low-nutrient sites at both ends of the gradient (fig. 4.11). Residence time is a function of two processes—leaf longevity and the proportion of nutrients resorbed from senescing leaves. Leaf longevity across the sites (calculated as leaf biomass divided by annual leaf litter-fall) ranged from less than two to almost five years, with the longest-lived leaves in the 0.3 ky site (fig. 4.11a). This inverse relationship between leaf longevity and nutrient availability is characteristic of woody perennials in general (Reich et al. 1992). The pattern for root longevity was not as clear; although the low-P 4100 ky site had very long-lived roots (~ 6.2 years), root longevity in the low-N 0.3 ky site (1.5 years) and the fertile 20 ky site (2.1 years) did not differ significantly (Ostertag 2001).

Nutrient resorption also varied substantially across the sites. Although there is no systematic, global relationship between nutrient availability and nutrient resorption (Aerts 1996, Killingbeck 1996), there is a strong negative correlation between nutrient availability and the proportion of N and P resorbed for *Metrosideros* on the Hawaiian age gradient (Riley and Vitousek 1995, Herbert and Fownes 1999) (fig. 4.11b). Combining leaf longevity and nutrient resorption, retention of N and P within the canopy of fertile intermediate-aged sites ranged from 3–5 years, while

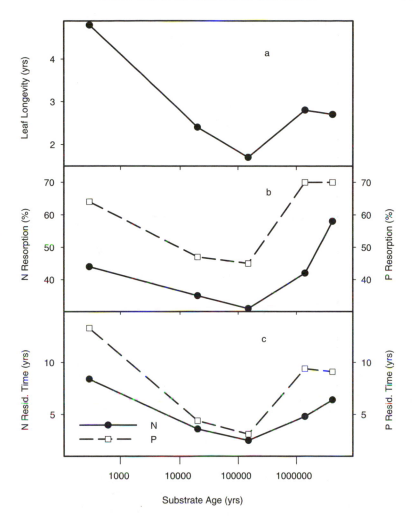

FIGURE 4.11. Leaf longevity, nutrient resorption, and nutrient residence times across the Hawaiian age gradient, from information in Herbert and Fownes (1999). (a) Mean leaf longevity, calculated as total leaf mass divided by annual leaf litterfall, is greatest in the young site and smallest in intermediate-aged sites. (b) Resorption of N and P prior to leaf senescence is greatest in the low-nutrient young and old sites. (c) Residence times of N and P in the canopy, calculated as mean leaf longevity corrected for resorption, peak in the young and old sites.

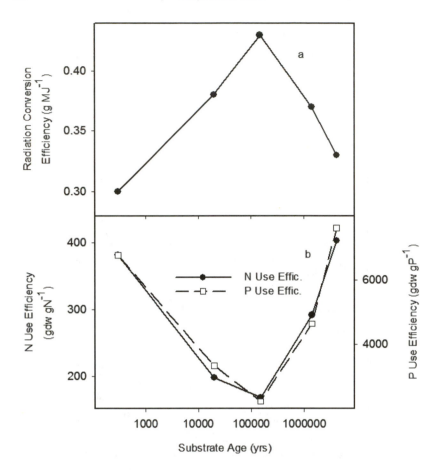

FIGURE 4.12. Radiation conversion efficiency and nutrient use efficiency of forest stands across the Hawaiian age gradient, modified from Herbert and Fownes (1999) and unpublished data on twig and wood chemistry. (a) Radiation conversion efficiency (e) is calculated as ANPP per unit of light absorbed in the canopy; it peaks in intermediate-aged sites. (b) N-use and P-use efficiencies are calculated as ANPP divided by the quantity of each nutrient in leaf and twig litterfall and annual wood increment. N and P efficiencies are the mirror image of radiation-conversion efficiency, suggesting that there is a tradeoff between radiation-conversion and nutrient-use efficiencies (Herbert and Fownes 1999).

that in the youngest and oldest sites ranged from 6.5 to 13 years (fig. 4.11c). Clearly these forests maintain productivity on low-nutrient sites in large part by increasing the residence time of nutrients within vegetation, thereby obtaining a higher NPP per unit of nutrient over the lifetime of a unit of nutrient in the canopy (Aerts and Chapin 2000). The overall variation in nutrient-use efficiency across the sites largely reflected variation in nutrient residence time (fig. 4.12a), with a substantial contribution from nutrient productivity only in the 4100 ky site.

Production per unit of light absorbed (light-use or canopy efficiency, ε) varied inversely with nutrient-use efficiency, peaking in the intermediate-aged sites (fig. 4.12b). Herbert and Fownes (1999) interpret this overall pattern as representing a tradeoff between light-use efficiency and nutrient-use efficiency, with nutrient-use efficiency greater in low-nutrient sites and light-use efficiency greater in high-nutrient sites. Proximately, the low ε of forests on infertile sites could reflect both lower maximum rates of leaf-level photosynthesis by low-nutrient leaves (Reich et al. 1997), and the preponderance of older, partially-shaded leaves in their canopies.

Decomposition and Nutrient Regeneration

How do differences along the gradient in nutrient-use efficiency (and so in plant chemistry) affect rates of decomposition and nutrient release? The most important controls over rates of litter decomposition are climate, the structural and chemical properties that control the inherent decomposability of litter itself (litter quality), and nutrient availability and the activity of decomposer organisms in the site where decomposition occurs (together termed "site quality") (Swift et al. 1979, Lavelle 1997). The Hawaiian age gradient was selected to minimize variation in climate, but both litter chemistry and nutrient availability change substantially along it. As discussed above, foliar N and P concentrations are low in the infertile young and old sites (fig. 4.6b), and this difference is accentuated by proportionally greater resorption of N and P during leaf senescence in these sites (fig. 4.12b). Consequently, concentrations of N and P in litter are much lower in young and old sites than in intermediate-aged sites (fig. 4.13)— enough so that the substrate quality for decomposition should be substantially lower in the young and old sites. In addition, low nutrient availability in the young and old sites could limit rates of decomposition directly and/ or constrain the types of decomposers that could be active therein.

A number of studies have measured rates of decomposition in several sites across the gradient (Crews et al. 1995, Thompson and Vitousek 1997, Ostertag and Hobbie 1999, Hobbie and Vitousek 2000). Several of these evaluated the importance of litter quality and/or site quality by measuring rates of decomposition of a single litter type at multiple sites (where any variation is due to site characteristics) and/or rates of decomposition of multiple litter types collected in a range of sites and decomposed in a single site (where any variation is due to characteristics of the litter itself). I term the former "site quality experiments" and the latter "litter quality experiments."

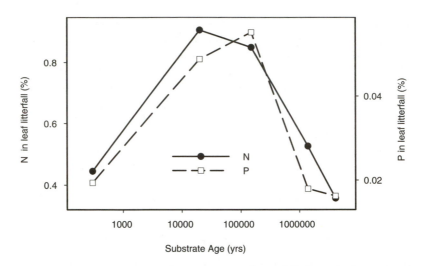

FIGURE 4.13. Concentrations of N and P in leaf litterfall along the Hawaiian age gradient, modified from Herbert and Fownes (1999).

Two studies measured decomposition rates of *Metrosideros* leaf litter in multiple sites across the age gradient; both included site quality and litter quality experiments. One used the 0.3 ky, 150 ky, and 4100 ky sites (Crews et al. 1995, Vitousek et al. 1997b), as well as two 20 ky sites ~ 300 meters higher and lower in elevation than the 20 ky site on the gradient. Litter chemistry in these higher and lower elevation sites is similar to that at 1200 m (Scowcroft et al. 2000, Hobbie and Vitousek 2000), and differences in rates of decomposition between these sites and 1200 m were adjusted for differences in temperature as described in Crews et al. (1995). The other study used the 0.3 ky, 20 ky, and 4100 ky sites (Hobbie and Vitousek 2000, Vitousek and Hobbie 2000). Independent collections of litter were made for each experiment, and inter-site differences in litter N and P concentrations were consistent across collections (table 4.2). Concentrations of lignin (and polyphenols—Hättenschwiler et al. 2003) were relatively high in all the sites, but lignin concentrations were inconsistent from time to time and particularly method to method (table 4.2). Rates of decomposition were summarized and compared across the experiments using the exponential decay constant k (Swift et al. 1979) to provide a standardized comparison of time-integrated rates of decomposition across sites and substrates.

The two studies of *Metrosideros* decomposition yielded remarkably similar results. Leaf litter decomposition is slow in the low-nutrient 0.3 ky and 4100 ky sites, and a factor of > 2 faster in the fertile 20 ky and 150 ky

TABLE 4.2

Initial chemistry of the litter used in decomposition experiments along the Hawaiian age gradient. All values in %.

Substrate	Element or Compound	0.3 ky	20 ky	150 ky	4100 ky	Reference
Metrosideros leaf litter	N	0.40	0.80	0.74	0.37	Crews et al. (1995)
	P	0.026	0.053	0.054	0.022	
	lignin[1]	26.0	36.0	25.0	37.0	
Metrosideros leaf litter	N	0.33	0.88	—	0.37	Hobbie & Vitousek (2000)
	P	0.022	0.051	—	0.021	
	lignin[2]	21.8	19.4	—	16.9	
	lignin[3]	40.2	40.7	—	41.9	
Fine roots (mostly *Metrosideros*)	N	0.42	0.42	—	0.30	Ostertag & Hobbie (1999)
	P	0.050	0.040	—	0.022	
	lignin[3]	26.9	30.4	—	32.7	
Cheirodendron trigynum leaf litter	N	0.40	0.61	0.52	0.48	Thompson & Vitousek (1997)
	P	0.031	0.039	0.044	0.039	
	lignin[1]	6.5	13.0	7.9	7.1	

Notes

[1]Lignin by acid detergent fiber (van Soest and Wine 1968).

[2]Lignin determined by acetyl bromide (Iiyama and Wallis 1990).

[3]Lignin by forest products method (Ryan et al. 1989); yields higher results than other methods when identical samples are analyzed.

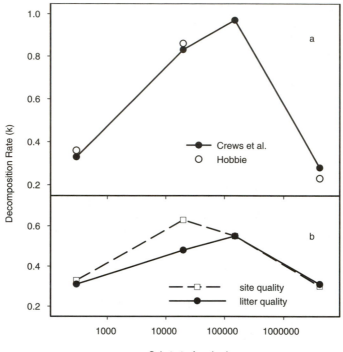

FIGURE 4.14. Decomposition of *Metrosideros* leaf litter across the substrate age gradient, from information in Crews et al. (1995), Vitousek et al. (1997b), and Hobbie and Vitousek (2000). (a) Two independent studies measured decomposition of litter in the sites where it was produced (solid symbols from Crews et al. 1995; hollow ones from Hobbie and Vitousek 2000). Both yield more rapid decomposition in high-nutrient intermediate-aged sites. (b) Decomposition of leaf litter that was produced in a single site and decomposed in each of the sites across the gradient, as a measure of site quality; and decomposition of litter collected in each site and decomposed in a single common site as a measure of litter quality. Both site and litter quality contribute to the greater rates of decomposition in intermediate-aged sites.

sites (fig. 4.14a)—even though climate did not differ among sites. Litter produced in the more fertile sites is inherently more decomposable than that from less-fertile sites; it decomposes significantly faster in a common site (fig. 4.14b). Similarly, a consistent type of litter decomposes more rapidly in the more fertile sites (fig. 4.14b), demonstrating that both site and litter quality contribute to the overall pattern of faster litter decomposition in more fertile sites.

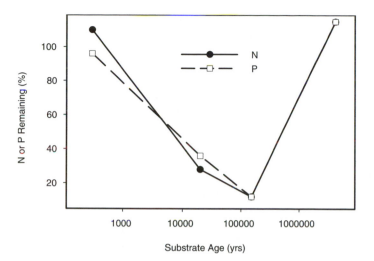

FIGURE 4.15. The fraction of initial N and P remaining after two years of *Metrosideros* leaf litter decomposition in the site where it was produced, modified from Vitousek et al. (1997b). In the intermediate-aged sites, a larger fraction of the greater quantity of nutrients in litter rapidly returns to biologically available forms. N and P remaining in litter can be greater than 100% of initial contents due to immobilization of N and P.

Decomposition of other substrates yielded broadly similar results (table 4.3). Leaf litter of the widespread native tree *Cheirodendron trigynum* collected and decomposed at the 20 ky site decomposed more rapidly than *Cheirodendron* collected and decomposed at the 0.3 ky site (*k* = 3.2 and 2.5 for the 20 ky and 0.3 ky sites respectively) (Thompson and Vitousek 1997). Roots (which could not be separated by species) decomposed at similar rates in the 0.3 ky and 20 ky sites, but significantly more slowly in the 4100 ky site (Ostertag and Hobbie 1999).

These differences in rates of litter decomposition affect the rates at which nutrients cycle through litter and are released into biologically available forms. For *Metrosideros* litter produced and decomposed in the low nutrient 0.3 ky and 4100 ky sites, the absolute quantity of N and P within litter increased due to microbial uptake and retention of N and P, so that decomposing litter was a sink rather than a source of nutrients for at least two years. In contrast, most of the N and P in litter produced and decomposed in the more fertile 20 ky and 150 ky sites had been released from litter and was back in circulation within two years (fig. 4.15).

TABLE 4.3

Decomposition constants (per year) for litter that was produced on the Hawaiian age gradient, and decomposed at the site where it was produced (in situ); for constant litter types that were decomposed in several of the sites (site quality experiment); and for litter that was produced in several age gradient sites and decomposed together in a common site (litter quality).

In Situ	0.3 ky	20 ky	150 ky	4100 ky	Note	Reference
Metrosideros leaf litter	0.33	0.83	0.97	0.28	1	Vitousek et al. 1997B
"	0.36	0.86	—	0.23	2	Hobbie & Vitousek 2000
"	0.34	0.56		0.18	3	Ostertag & Hobbie 1999
Roots (mostly *Metrosideros*)	0.54	0.51	—	0.30	4	Ostertag & Hobbie 1999
Cheirodendron leaf litter	2.5	3.2	—	—	5	Thompson & Vitousek 1997

Site Quality	0.3 ky	20 ky	150 ky	4100 ky	Note	Reference
Low-nutrient *Metrosideros* leaf litter	0.33	0.63	0.55	0.30	6	Vitousek et al. 1997b
High-nutrient *Metrosideros* leaf litter	0.42	0.86		0.48	7	Hobbie & Vitousek 2000

Litter Quality

	0.3 ky	20 ky	150 ky	4100 ky	Note	Reference
Metrosideros leaf litter	0.31	0.48	0.55	0.31	8	Vitousek et al. 1997b
Roots (mostly *Metrosideros*)	0.53	0.63	—	0.33	9	Ostertag & Hobbie 1999
Cheirodendron leaf litter	2.5	2.5	2.9	2.9	10	Thompson & Vitousek 1997

Notes

[1]Coarse-mesh litter bags; 20 ky result calculated from measurements at higher and lower elevations.
[2]Coarse-mesh litter bags.
[3]Fine-mesh litter bags, for comparison with root decomposition.
[4]Fine-mesh bags.
[5]Coarse-mesh bags.
[6]Low-quality substrate from 1855 Mauna Loa flow; otherwise as in #1 above.
[7]High-quality substrate from 20 ky site; otherwise as in #2 above.
[8]Decomposed in infertile 0.3 ky site; otherwise as in #1 above.
[9]Decomposed in moderately infertile 2.1 ky site; otherwise as in #4 above.
[10]Decomposed in infertile 0.3 ky site; otherwise as in #5 above.

Soil Organic Matter Turnover

Soil organic matter (SOM) is the dispersed, microbially processed, and mostly recalcitrant residue of litter decomposition; it consists of a wide variety of different compounds with very different rates of decomposition. Models of SOM turnover often divide it conceptually into three pools: an active, decomposable fraction (< 5% of the total); a slow fraction that turns over in years to decades; and a passive fraction with turnover times of more than a century (Parton et al. 1993). Short-term measurements of SOM decomposition (e.g., CO_2 evolution) mostly reflect the dynamics of the most labile fraction; however, longer-term soil C storage and nutrient supply are controlled by the much larger slow pool of SOM (Townsend et al. 1995, Trumbore et al. 1996).

Torn et al. (submitted) estimated rates of decomposition of soil organic matter (SOM) in the forest floor and A-horizon of sites across the Hawaiian age gradient using two methods. One approach used long-term incubations of soil in the laboratory (Townsend et al. 1997), following rates of CO_2 evolution through an initial pulse to a lower, more or less stable rate that is assumed to reflect turnover of the slow SOM pool. The other method used the substantial pulse input of ^{14}C into the atmosphere derived from nuclear testing in the 1950s and early 1960s. The magnitude and timing of this ^{14}C pulse are well-defined—and given a number of assumptions, the current abundance of ^{14}C in SOM can be used to calculate SOM turnover (Trumbore 2000).

Torn et al. (submitted) calculated that 70–90 percent of the SOM in the forest floor and A horizon is in the slow fraction across all the sites. Turnover times of this fraction fell in the range of 10–100 years using both techniques in both horizons. Both approaches yielded similar patterns across the sites, although decomposition rates calculated using ^{14}C were slower than incubation-based estimates. Turnover of both forest floor and A-horizon SOM was slowest in the 2.1 and 4100 ky sites and fastest in the 150 and 1400 ky sites, for both methods and both horizons (fig. 4.16a, b) (Torn et al. submitted). The pattern of more rapid SOM decomposition in surface horizons of the more-fertile sites does not extend to the complete inventory of SOM (to one meter depth) in these sites; most of the deeper SOM is in mineral-associated passive pools with mean ^{14}C ages > 20,000 yrs (Torn et al. 1997).

PLANT-SOIL-MICROBIAL FEEDBACKS

The patterns in soil N and P availability along the Hawaiian age gradient are caused (ultimately) by the geochemical processes described by Walker and Syers (1976). The youngest sites are limited by N because N accu-

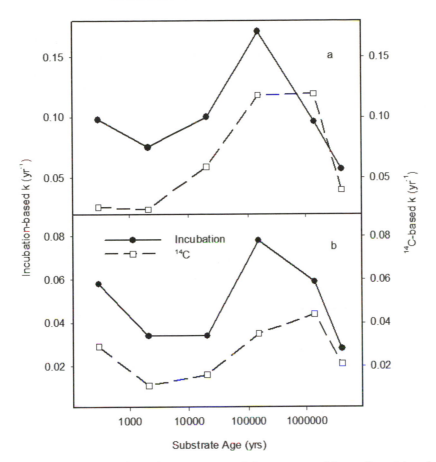

FIGURE 4.16. Two independent measures of the turnover of forest floor (a) and A-horizon soil organic matter (b) along the Hawaiian age gradient, calculated from information in Torn et al. (submitted). Turnover was calculated based on the release of CO_2–C during long-term incubations (solid lines) and modeled based on the quantity of[14] C from atmospheric nuclear bomb testing that remains in organic matter (dashed lines). The methods yield different absolute values but very similar patterns; by both methods, the turnover of forest floor peaks in intermediate-aged sites. Turnover of A-horizon organic matter was high in both the youngest site and intermediate-aged sites.

mulates slowly from the atmosphere, and they are low in P because most P remains bound in primary minerals in the soil. The oldest site is limited by P because most of the P that has been supplied by basalt weathering during its history has been lost, and the majority of the remainder is bound into insoluble and physically protected forms in soil. However, biological responses to a geochemically driven shortage of nutrients can further reduce

the availability of nutrients, setting in motion a positive feedback that reinforces nutrient deficiency in infertile sites and greater nutrient availability in fertile sites. The potential for such a plant-soil-microbial positive
feedback has been described in a number of ecosystems (Vitousek 1982,
Pastor and Post 1986, Wedin and Tilman 1990, Hobbie 1992, Binkley
and Giardina 1998, Aber and Melillo 2001—although see Tateno and
Chapin 1997), but the components and controls of the feedback can be
demonstrated and tested unusually clearly in Hawaiian ecosystems.

The basic features of the plant-soil-microbial feedback are outlined in
fig. 4.17. Trees growing in the low-nutrient young and old sites on the
substrate age gradient (fig. 4.4b, 4.5) use nutrients more efficiently than
those in the relatively rich intermediate-aged sites, in effect trading a lower

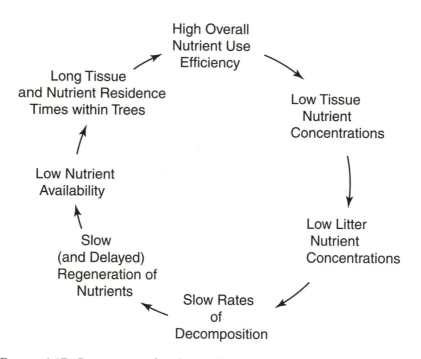

FIGURE 4.17. Components of a plant-soil microbial positive feedback. In comparison to *Metrosideros* forests on relatively fertile intermediate-aged sites, *Metrosideros* forests on young and old sites on the Hawaiian substrate age gradient
experience low N and/or P availability. These forests retain nutrients in their
canopies longer, have higher nutrient-use efficiencies, lower tissue and litter nutrient concentrations, and slower rates of litter decomposition and nutrient regeneration. The overall effect is to slow rates of nutrient cycling in infertile
sites—further reducing nutrient availability.

radiation conversion efficiency (ε) for a greater nutrient use efficiency (fig. 4.12). They do so primarily by increasing the residence time of nutrients in the canopy, increasing both leaf longevity and the resorption of nutrients from senescing tissues (fig. 4.11). As a consequence, both leaf (fig. 4.6) and litter (fig. 4.13) nutrient concentrations are low. These low-nutrient tissues decompose more slowly than those produced in more fertile intermediate-aged sites (fig. 4.14), due both to characteristics of the litter itself and of the sites where decomposition occurs. Moreover, the low-nutrient litter of the infertile young and old sites immobilizes nutrients longer during decomposition, causing nutrient release to lag behind mass loss in these sites (fig. 4.15). Even soil organic matter turns over more slowly in these infertile sites (fig. 4.16). Overall, the effect of the feedback is to slow rates of nutrient cycling and hence reduce nutrient availability within infertile sites, and conversely to accelerate nutrient cycling in richer sites (fig. 4.17).

This brief illustration of plant-soil-microbial feedback describes our observations along the Hawaiian substrate age gradient—but it does not explain why the feedback works as it does. For example, decomposition of low-nutrient litter could be relatively slow because trees chemically defend long-lived leaves against grazers and pathogens, and those defensive compounds also inhibit decomposers (Rhoades 1979); because the rate of decomposition itself is limited by nutrient supply; and/or for other reasons. In the next chapter, I use the results of long-term fertilization experiments to evaluate how plant-soil-microbial feedback can be initiated and regulated in Hawaiian forests, and in the last chapter I will explore how nutrient input/output budgets can constrain—or sustain—the importance of this feedback in the long term.

Chapter Five

EXPERIMENTAL STUDIES OF NUTRIENT

LIMITATION AND THE REGULATION

OF NUTRIENT CYCLING

GRADIENT STUDIES and other cross-ecosystems comparisons can be used to identify patterns in nutrient cycling and suggest possible controlling processes. However, testing the importance of those processes and controls generally requires experimentation. Along the Hawaiian age gradient, forests differ substantially and predictably in nutrient availability, plant chemistry, decomposition, and overall nutrient cycling (Crews et al. 1995, chapter 4); these patterns suggest that nutrient supply limits productivity and other ecosystem processes in the infertile youngest and oldest sites on the gradient, and that nutrient availability, plant nutrient use efficiency, and decomposition interact in a positive feedback that slows nutrient cycling in poor sites and speeds it in nutrient-rich sites (fig. 4.17).

In this chapter, I make use of several fertilization experiments in Hawaiian forests to ask:

1. Are NPP and other ecosystem processes in the infertile young and old sites on the Hawaiian age gradient indeed limited by nutrient supply? If so, is N limiting on young substrates and P limiting on old substrates, as would be expected from the Walker and Syers (1976) model?
2. Can changes in nutrient availability drive a plant-soil-microbial positive feedback that exacerbates nutrient limitation in low-nutrient sites? If so, what are the limits to this feedback?

To answer question 1, first I need to say what I mean—and do not mean—by nutrient limitation. I define nutrient limitation as occurring when the addition of an essential element increases the growth of individual organisms or populations, or increases the rate of a biological process (Chapin et al. 1986). The existence of nutrient limitation can be inferred from measurements of soil nutrient availability, foliar nutrient concentrations or element ratios, and/or root uptake capacity for particular nutrients (van den Driessche 1974, Harrison and Helliwell 1979, Powers 1980, Aerts and Chapin 2000); most of these indices predict nutrient limitation reasonably well, particularly within a restricted range of forests where they have been calibrated against responses to fertilization. Ultimately, however, nutrient limitation is defined operationally and must be demonstrated empirically.

Second, the demonstration that a particular nutrient limits an organism or process does not preclude the possibility of simultaneous or sequential limitation by other resources. Rather, multiple resource limitation by light, water, CO_2, and one or more soil-derived nutrients probably represents the normal situation for terrestrial plants (Field et al. 1992, Rastetter et al. 1997). On a physiological level, adding N can allow plants to produce more RUBP carboxylase enzyme and photosynthetic machinery, and adding water can allow them to keep their stomates open wider and/or longer—both of which increase CO_2 fixation. Similarly, an increase in CO_2 concentration increases both water- and nitrogen-use efficiency. Each of these resources can be limiting, in the sense that adding it alone allows higher rates of photosynthesis and the potential for greater growth.

In addition to demonstrating nutrient limitation, fertilization experiments can yield insight into the mechanisms controlling nutrient cycling in ecosystems. If the differences in nutrient cycling that we observe along the Hawaiian age gradient are caused primarily by differences in nutrient availability, then adding the limiting nutrient(s) to low nutrient sites should make them function like sites that are naturally higher in nutrients. If they do so, then the processes and feedbacks involved can be traced and understood. If nutrient additions do not make low nutrient systems function more like high-nutrient systems—even though all of the systems are dominated by the same tree species, and have much else in common—then either nutrient availability does not drive the observed differences among sites, or there are additional causes of these differences that I would like to try to understand.

I will focus here on nutrient limitation and components of the plant-soil-microbial feedback outlined in fig. 4.17, but we also have used the longer-term fertilizer experiments to evaluate: (1) N losses via leaching and trace gas emissions following short- and long-term fertilization in N-limited and "N-saturated" forest ecosystems (Hall and Matson 1999, 2003; Lohse 2002); (2) controls of biological N fixation by epiphytes and by litter heterotrophs (Vitousek 1999, Crews et al. 2000, Vitousek and Hobbie 2000); and (3) the integration of leaf, stand, and ecosystem processes in N versus P limited forest ecosystems (Cordell and Goldstein 1999; Cordell et al. 2001a, b; Hobbie and Vitousek 2000; Harrington et al. 2001).

Fertilization Experiments

We have carried out six factorial fertilization experiments in Hawaiian montane forests. Three are on the substrate age gradient, spanning the range of sites from young, infertile, and putatively N-limited in the 0.3 ky site; through intermediate-aged and relatively fertile at 20 ky; to old, infertile, and putatively P-limited at 4100 ky. Nutrient additions were initiated in 1985, 1993, and 1991 in the 0.3 ky, 20 ky, and 4100 ky sites respectively, and they continue through the present. The other three fertilization experiments are:

- a 26-year-old tephra deposit (0.026 ky), in the Byron's Ledge area of Kīlauea Volcano. This site is slightly drier than those on the substrate age sequence (2200 mm/yr vs. 2500) (Vitousek et al. 1993); it was fertilized in part to provide a context for ecosystem-level effects of an invading symbiotic N-fixer (Vitousek and Walker 1989).
- an 1855 pāhoehoe and an 1852 'a'ā lava flow on Mauna Loa. These sites are relatively wet (4400 mm yr^{-1}) (Raich et al. 1996); they were fertilized in part to evaluate how the texture of parent material influences nutrient limitation in developing ecosystems.

All six experiments used replicated complete-factorial designs, with main factors of: N, applied at 100 kg ha^{-1} yr^{-1}; P, also at 100 kg ha^{-1} yr^{-1}; and a mixed nutrient treatment that we term "T" that incorporates all other elements essential to plant growth (other than N and P). The quantities and forms of elements applied are summarized in table 5.1. In total, there were eight treatments (control, +N, +P, +T, +NP, +NT, +PT, and

TABLE 5.1

Quantities and forms of elements applied in the fertilization experiments. All of the experiments were fully factorial, using N, P, and all other essential elements combined (abbreviated T) as main treatments.

Treatment	Element(s)	Form	Application rate (kg ha^{-1} yr^{-1})
N	N	NH_4NO_3, urea	100
P	P	Triple super P	100
T	Ca	Dolomite	100
	Mg	Dolomite	58
	K	K_2SO_4	100
	S	K_2SO_4	40
	Fe	Granusol #2gb5	8
	Mn	"	8
	Zn	"	8
	Cu	"	2.25
	B	"	0.75
	Mo	Supplement	0.01

TABLE 5.2

Characteristics of factorial fertilization experiments along the substrate age gradient, and elsewhere in the Hawaiian Islands.

Site	Age	Substrate	Initiated	Completed	Plots	Replicates/ treatment	Design	Reference
Thurston	0.3 ky	tephra	1985	ongoing	15 × 15 m	4	completely randomized	Vitousek et al. 1993
Laupāhoehoe	20 ky	tephra	1993	ongoing	tree-centered 5 m radius	6	randomized block	Vitousek & Farrington 1997
Kōke'e	4100 ky	?[1]	1991	ongoing	15 × 15 m	4	randomized block	Herbert & Fownes 1995
Byron's Ledge	0.026 ky	tephra	1985	1987	tree-centered 2 m radius	5	completely randomized	Vitousek et al. 1993
1855 flow	0.136 ky	pāhoehoe	1991	1995	10 × 10 m	4	randomized block	Raich et al. 1996
1852 flow	0.139 ky	'a'ā	1991	1995	10 × 10 m	4	randomized block	Raich et al. 1996

Note
[1] The soil at Kōke'e is so highly weathered that the original parent material (tephra or lava flow) cannot be identified, but if the Kaua'i volcano erupted similarly to younger ones, the site was tephra-covered early in its development.

+NPT), replicated four to six times per site. Variations in plot sizes and layouts, details of experimental design, and the timing of experiments are summarized in table 5.2. In all of the experiments (except in the 4100 ky site, as discussed below) we identified nutrient limitation by determining growth responses following two years of fertilization. Several years later, we evaluated how nutrient cycling and components of plant-soil-microbial feedback responded to fertilization in the 0.3 ky and 4100 ky sites.

Nutrient Limitation

Overall, responses to fertilization strongly support predictions based on the Walker and Syers (1976) conceptual model. The model suggests that N should be in short supply in young sites, and additions of N to the 0.3 ky site significantly increased tree diameter increment and annual litterfall by the second year of fertilization (fig. 5.1a, table 5.3) (Vitousek et al. 1993). The P and T treatments did not affect growth or litterfall, either alone or interactively with N. The height growth and frond production of understory tree ferns (*Cibotium glaucum*) was increased by added N;

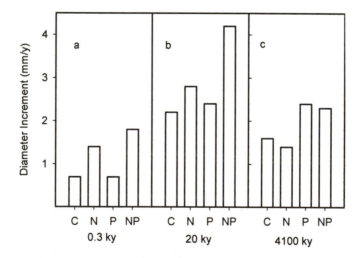

FIGURE 5.1. Tree diameter increments after two years of fertilization (18 months in the oldest site) in fertilized and control plots of (a) the 0.3 ky (Vitousek et al. 1993), (b) 20 ky (Vitousek and Farrington 1997), and (c) 4100 ky sites (Herbert and Fownes 1995). N additions increased growth significantly in the young site; P did so in the old site, and neither N nor P alone increased growth in the intermediate-aged site. Additional plots were fertilized with essential nutrients other than N and P; these did not stimulate growth anywhere.

added P also increased height growth of tree ferns, but T had no effect (Walker and Aplet 1994).

Results from the three additional fertilizer experiments on young sites also supported Walker and Syers' model. In the 0.026 ky site, added N increased *Metrosideros* diameter growth (table 5.3), height growth, leaf production, and leaf-level photosynthesis significantly, and decreased leaf longevity (Vitousek et al. 1993); the P and T treatments yielded no significant effects or interactions.

The experiments on Mauna Loa lava flows should provide a stronger challenge to Walker and Syers, because lava flow substrates are substantially coarser-textured than is tephra, with pāhoehoe in particular presenting much less surface area accessible to weathering (Raich et al. 1996). Consequently, the supply of P and other rock-derived nutrients should be much smaller on lava flows than on tephra. We found that tree diameter (table 5.3) and height growth, and growth of the understory mat-forming fern *Dicranopteris linearis*, responded to additions of both N and P (individually and in combination) in both 'a'ā and pāhoehoe sites (Raich et al. 1996); the T treatment had no significant effects. However, the response to added N far exceeded that to P, even on pāhoehoe lava. Overall, N limitation to plant growth is paramount in all of the young sites.

In the 4100 ky site, the Walker and Syers (1976) model suggests P limitation, and in fact P additions significantly increased tree diameter increment (fig. 5.1c), litterfall (table 5.3), leaf area index (LAI), and aboveground NPP (ANPP) within eighteen months following the initiation of fertilization there (Herbert and Fownes 1995). Additions of N also increased litterfall significantly, and there was a significant negative P × T interaction for diameter increment. Darrell Herbert, who developed and initiated this fertilization experiment, intended to assess responses to nutrient additions after two years (as in the other sites). However, Hurricane Iniki passed directly over the site eighteen months into the experiment and altered that plan. (Eighteen-month measurements were collected the day before the hurricane struck.) The hurricane reduced LAI substantially, especially in P-fertilized plots (Herbert et al. 1999), but it snapped, tipped, or otherwise killed few trees. Fertilizer additions continued after the hurricane; by the second post-hurricane year, the stimulatory effects of P additions to LAI, diameter increment, and litterfall were again highly significant (Herbert et al. 1999).

The Walker and Syers model suggests that the availability of N and P should equilibrate at a relatively high level in ecosystems on intermediate-aged substrates, and I anticipated that the relatively fertile 20 ky site would not respond to nutrient additions. In fact, while diameter growth did not increase following additions of N or P (or T) alone, adding N and

TABLE 5.3

Fertilization responses after two years (18 months in the 4100 ky site) in factorial experiments on the substrate age gradient, and at other Hawaiian sites. The T treatment (all nutrients other than N and P) rarely affected plant growth or related measures, and +T and −T (e.g., N and N+T) results are combined here. Diameter measurements are reported in mm/yr, litterfall in g m^{-2}yr^{-1}, and foliar N and P in %. References as in table 5.2.

Site	Response	Control	+N	+P	+N and P
(0.025 ky)	diameter increment	4.6	13.3	6.6	10.6
	foliar N	0.83	0.92	0.83	1.10
	foliar P	0.095	0.072	0.14	0.11
0.3 ky	diameter increment	0.7	1.4	0.7	1.8
	litterfall	534	580	470	690
	foliar N	0.73	0.74	0.70	0.78
	foliar P	0.042	0.044	0.068	0.050
20 ky	diameter increment	2.2	2.8	2.4	4.2
	foliar N	1.17	1.34	1.27	1.33
	foliar P	0.090	0.10	0.11	0.12
4100 ky	diameter increment	1.6	1.4	2.4	2.3
	litterfall	453	526	514	634
	foliar N	0.80	0.87	0.85	0.94
	foliar P	0.058	0.061	0.10	0.12
1855 lava flow	diameter increment	0.5	1.8	1.3	3.9
	foliar N	0.71	0.80	0.69	0.90
	foliar P	0.073	0.066	0.15	0.14
1852 lava flow	diameter increment	0.7	3.2	1.4	5.4
	foliar N	0.62	0.73	0.67	0.77
	foliar P	0.068	0.057	0.16	0.097

P together caused a significant increase in growth (fig. 5.1b) (Vitousek and Farrington 1997). This result suggests that while the supply of N and P indeed equilibrated at a relatively high level in this intermediate-aged site, the trees could still respond to additional N and P as long as they received both. Overall, the pattern of growth responses to two years of nutrient additions is consistent with Walker and Syers' model—N supply limits growth in all four young sites, N and P supply have equilibrated at a relatively high level in the intermediate-aged site, P limits growth in the

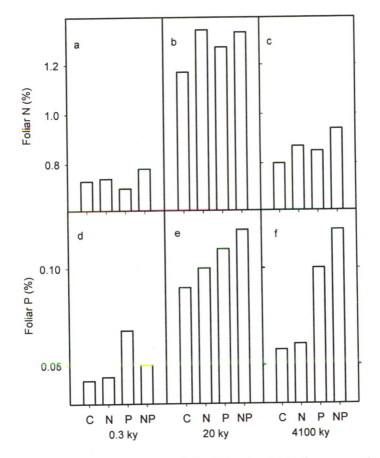

FIGURE 5.2. Effects of fertilization on foliar N (a–c) and P (d–f) concentrations in the 0.3 ky, 20 ky, and 4100 ky sites, from Vitousek et al. (1993), Vitousek and Farrington (1997), and Herbert and Fownes (1995), respectively. Foliar P concentrations were more responsive to P additions than were foliar N concentrations to N additions, especially where P limits plant growth.

oldest site, and other nutrients (the T treatment) have little or no effect on productivity.

In addition to tree growth, we determined foliar nutrient concentrations two years post-fertilization (18 months at the 4100 ky site) in all treatments of all six experiments. Fertilization with N increased foliar N concentrations significantly everywhere, but the increases were not spectacular anywhere; additions of N alone increased foliar N less than 15% (fig. 5.2a–c, table 5.3). In contrast, P additions increased foliar P strikingly in all of the sites (fig. 5.2d–f, table 5.3)—particularly at the sites where P additions increased growth, the 4100 ky site on the substrate age gradient and the two sites on young Mauna Loa lava flows, where foliar P nearly or fully doubled following fertilization (table 5.3).

NUTRIENT AVAILABILITY AND PLANT-SOIL-MICROBIAL FEEDBACK

Several years later, we evaluated controls of nutrient cycling and plant-soil-microbial feedbacks in fertilized plots of the N-limited 0.3 ky site and the P-limited 4100 ky site to test whether fertilization with the limiting nutrient(s) makes forests on infertile soils function more like those on fertile soils (fig. 4.17). A few parallel measurements were undertaken in fertilized plots of the relatively fertile 20 ky site. We waited until most tissues in the fertilized plots had been produced with an elevated nutrient supply, so that patterns of C and nutrient allocation could equilibrate to a greater nutrient supply.

Tissue Nutrient Concentrations

The accumulation of four (4100 ky site) and ten (0.3 ky site) more years of fertilization led to a small increase in foliar N, relative to two years post-fertilization—and a large increase in foliar P, especially in the P-limited 4100 ky site (table 5.4). Although foliar N concentrations in N-fertilized plots of the N-limited 0.3 ky site never approached those in unfertilized trees of the 20 ky site, foliar P concentrations in P-fertilized plots of the P-limited 4100 ky site greatly exceeded those in more fertile sites (fig. 4.6, table 5.4). Nutrient concentrations in roots and wood responded similarly to those in leaves (table 5.4); while root N concentrations increased nearly 50% following N fertilization in the N-limited site, root P concentrations responded more than 4-fold in the P-limited site.

Productivity

Longer-term responses of net primary production (NPP) to N and P fertilization confirmed the shorter-term measurements of plant growth reported in fig. 5.1, demonstrating that N limits productivity in the young

TABLE 5.4

Nutrient concentrations (%) in leaves, stemwood, and roots following
long-term fertilization in the 0.3 ky and 4100 ky sites on the Hawaiian
age gradient. ND = not determined. From Harrington et al. (2001).

		Thurston (0.3 ky)				Kōkeʻe (4100 ky)			
		Control	N	P	NP	Control	N	P	NP
N	Leaves	0.73	0.85	0.66	0.78	0.95	1.05	0.82	1.16
N	Wood	0.10	0.13	0.10	0.13	0.07	0.09	0.07	0.14
N	Roots	0.42	0.61	0.40	ND	0.30	0.44	0.54	ND
P	Leaves	0.058	0.054	0.095	0.077	0.055	0.067	0.247	0.188
P	Wood	0.010	0.008	0.028	0.020	0.006	0.007	0.023	0.034
P	Roots	0.050	0.036	0.113	ND	0.022	0.021	0.091	ND

site, and P does so in the old site. Aboveground NPP (ANPP) in the 0.3 ky
site increased from 650 g m^{-2} yr^{-1} in control plots to 1140 g m^{-2} yr^{-1} in
N fertilized plots; ANPP in N + P plots averaged 1635 g m^{-2} yr^{-1} (fig.
5.3a) (Harrington et al. 2001). Similarly, P additions to the 4100 ky site
increased ANPP from 950 g m^{-2} yr^{-1} to 1450 g m^{-2} yr^{-1}; N + P caused
a further increase to 1600 g m^{-2} yr^{-1} (fig 5.3b). Although BNPP did not
change with nutrient additions in either site, Ostertag (2001) observed a
significant increase in root turnover in P fertilized plots at the 4100 ky
site (fig. 5.3b). P additions affected some components of forest growth in
the young site, either alone or interactively with N; N additions had similar effects in the old site—suggesting that neither site is too far from co-
limitation by N and P.

Resource Efficiencies

LIGHT CAPTURE AND CONVERSION

Added nutrients can increase forest productivity if they allow forest
canopies to use incoming solar radiation more effectively. Forests can do
so in two ways—either by producing a greater leaf area per unit of
ground surface (leaf area index, or LAI) and capturing a greater fraction
of incoming solar radiation; or by utilizing the light that they do intercept
more efficiently, increasing NPP per unit of light absorbed (radiation conversion efficiency, or ε, described in chapter 4).

Long-term fertilization with the limiting nutrient increased LAI in both
sites, significantly so with P additions to the P-limited 4100 ky site (fig. 5.4b)

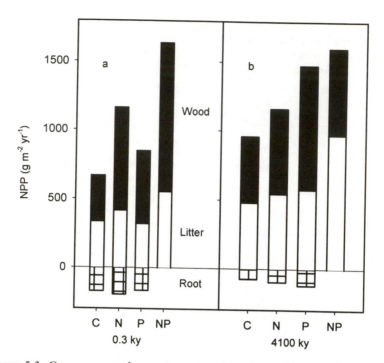

FIGURE 5.3. Components of net primary production (NPP) following 12 and 6 years of fertilization in the 0.3 ky (a) and the 4100 ky (b) sites, respectively, from Harrington et al. (2001). The hatched portion below 0 represents root production, from Ostertag (2001); root production was not measured in the N+P plots. The white portion of each bar represents annual leaf plus twig litterfall and the black portion represents annual wood increment.

(Harrington et al. 2001). Although statistically significant, the increased LAI corresponds to an increase in light absorption by the forest canopy of < 10% (Greg Asner, personal communication). Nutrient additions had a much larger effect on the efficiency with which forests made use of the light they captured; N additions to the N-limited site increased ε by 78%, and P additions to the P-limited site increased ε by 44% (fig. 5.4c,d) (Harrington et al. 2001). Susan Cordell also observed increased photosynthesis per unit of leaf area (A) in canopy leaves from P-fertilized plots in the 4100 ky site, a leaf-level parallel to the canopy-level increase in ε. However, there was no increase in A following N additions in the 0.3 ky site (Cordell et al. 2001a). P additions to the N-limited site and N additions to the P-limited site both increased ε (and A in the 4100 ky site), although not to the extent that the "main" limiting nutrient did (fig. 5.4c,d). Just as we observed along the gradient (chapter 4), differences in ε are

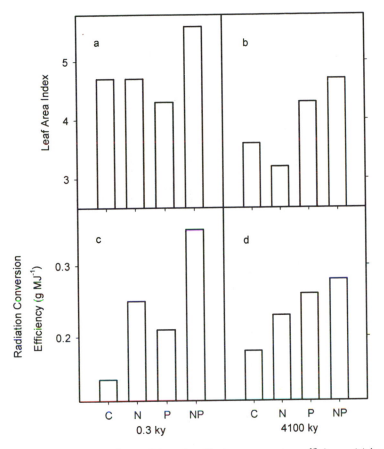

FIGURE 5.4. Leaf area index (LAI) and radiation conversion efficiency (ε) in fertilized and control plots of the 0.3 ky (a, c) and 4100 ky (b, d) sites, modified from Harrington et al. (2001).

more important than differences in LAI in determining how productivity responds to nutrient availability. In this sense, long-term fertilization clearly makes infertile ecosystems function more like naturally fertile ones.

NUTRIENT USE

Overall P-use efficiency (P nutrient-use efficiency, or P-NUE) decreased substantially following long-term P fertilization of the P-limited 4100 ky site (fig. 5.5d) (Harrington et al. 2001), in parallel to the lower nutrient-use efficiency in fertile intermediate-aged sites on the age gradient (fig. 4.12) (Herbert and Fownes 1999). However, N additions to the young N-limited site *increased* overall N-use efficiency (N-NUE) relative to that in unfertilized

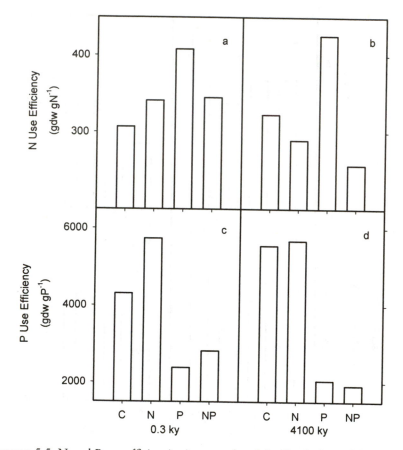

FIGURE 5.5. N and P use efficiencies in control and fertilized plots of the 0.3 ky and 4100 ky sites, from information in Harrington et al. (2001). (a, b) N-use efficiency increases following N fertilization of the N-limited 0.3 ky site. (c, d) P-use efficiency decreases following P fertilization of the P-limited 4100 ky site.

controls (fig. 5.5a). As discussed in chapter 4, the components of overall nutrient-use efficiency are nutrient productivity and nutrient residence time (Berendse and Aerts 1987). Consistent with the pattern observed along the substrate age gradient, long-term fertilization with the limiting nutrient decreased residence times for that nutrient in both sites (table 5.5). Additions of P to the 4100 ky site also decreased P productivity significantly, but additions of N to the N-limited 0.3 ky site *increased* N productivity (table 5.5)—reflecting a substantial increase in ANPP without a similarly large increase in foliar N concentrations (fig. 5.3, table 5.4). The combined effect of doubling N productivity with a smaller decrease

TABLE 5.5

Stand-level responses of canopy characteristics to long-term fertilization in the 0.3 ky (Thurston) and the 4100 ky (Kōke'e) sites on the substrate age gradient. N and P productivities are in grams dry weight per gram of nutrient in the canopy, leaf longevity and nutrient residence times are in yrs, and nutrient translocations are in %.

	Thurston			Kōke'e			Reference
	Control	N	P	NP	Control	N	P	NP	
N productivity	102	176	219	234	163	206	249	201	Harrington et al. (2001)
P productivity	1290	2910	1580	2430	2810	3270	860	1250	Harrington et al. (2001)
Leaf longevity	4.1	3.2	3.1	2.5	1.8	1.4	2.2	1.3	Harrington et al. (2001)
N residence time	10.5	9.4	7.6	6.8	6.0	4.0	6.9	4.8	calculated from above
P residence time	13.2	10.7	9.7	8.6	5.8	5.4	3.5	2.8	calculated from above
N retranslocation	61	66	59	63	70	65	68	73	Vitousek (1998)
P retranslocation	69	70	68	71	69	74	38	53	Vitousek (1998)

in N residence time is a small increase in the N-NUE of N-fertilized plots—contrary to what we observed in naturally more fertile sites on the substrate age gradient (fig. 4.12).

Decomposition

Both leaf litter and soil organic matter decompose more rapidly in more-fertile sites on the age gradient (figs. 4.14a, 4.16); for leaf litter, both the inherent decomposability of the litter and properties of the sites themselves contribute to this pattern (fig. 4.14b). This apparent sensitivity of decomposition to nutrient availability contributes substantially to plant-soil-microbial feedback (fig. 4.17). Fertilization experiments provide the opportunity to determine whether nutrient supply actually controls rates of decomposition—and if so, whether it does so directly by stimulating decomposition, indirectly through its effects on the decomposability of litter, or by some other pathway.

Several studies have evaluated decomposition following fertilization in Hawaiian montane forests. In one, I evaluated the litter quality (decomposability) of leaf litter produced in five of the fertilization experiments described above (Vitousek 1998). Concentrations of N in leaf litter were increased slightly (and generally significantly) by added N, while concentrations of P were increased much more substantially by added P, especially in sites where NPP was limited in whole or part by P (table 5.6). Added P also caused slight but significant decreases in lignin concentrations in four of the five sites (table 5.6), and concentrations of soluble polyphenols similarly were reduced by P fertilization in the 4100 ky site (Hättenschwiler et al. 2003). Litter produced in P-fertilized plots of the P-limited 4100 ky site (and the two Mauna Loa sites) decomposed significantly more rapidly than control-plot litter, with a positive N × P interaction. However, litter from N-fertilized plots in the N-limited 0.3 ky site did not decompose more rapidly than controls in a common site (table 5.6).

In a separate study, Sarah Hobbie used a reciprocal transplant design to evaluate fertilization effects on both litter and site quality for the three experiments on the age gradient. She determined the overall effect of fertilization by decomposing leaf litter in the fertilized and control plots where it was produced. The contribution of litter quality (decomposability) to this overall effect was assessed by decomposing litter from fertilized and control plots in the control plots of each site, and the direct effects of nutrient availability were assessed by decomposing litter from control plots in the control and fertilized plots of each site (Hobbie and Vitousek 2000). For litter (substrate) quality, she found that leaf litter from P-fertilized plots of the 4100 ky site decomposed more rapidly than

TABLE 5.6

The chemistry (in %) and decomposability (in % mass loss in one year) of
Metrosideros leaf litter produced in fertilization experiments on the substrate
age gradient and elsewhere in the Hawaiian Islands. Litter from all sites and
treatments was decomposed at a single site. From Vitousek (1998).

Experiment	Measure	Control	+N	+P	+N & P
0.3 ky	Foliar N	0.36	0.36	0.32	0.38
	Foliar P	0.024	0.021	0.042	0.030
	Lignin	21.5	19.8	22.4	21.2
	Decomposition	38	33	34	36
20 ky	Foliar N	0.73	0.81	0.82	0.85
	Foliar P	0.046	0.049	0.061	0.062
	Lignin	26.6	27.6	22.2	28.0
	Decomposition	40	40	42	41
4100 ky	Foliar N	0.30	0.37	0.31	0.32
	Foliar P	0.020	0.018	0.144	0.070
	Lignin	19.4	18.9	17.4	14.9
	Decomposition	32	31	37	42
1855 lava flow	Foliar N	0.28	0.35	0.32	0.36
	Foliar P	0.022	0.020	0.256	0.083
	Lignin	18.9	21.4	18.2	19.0
	Decomposition	35	37	43	44
1852 lava flow	Foliar N	0.29	0.36	0.29	0.39
	Foliar P	0.030	0.025	0.220	0.072
	Lignin	17.8	20.3	17.5	17.5
	Decomposition	35	37	36	46

control litter there, but N fertilization did not affect litter decomposability significantly in either site (fig. 5.6c,d). Measurements of the direct effects of nutrient additions (changes in site quality) yielded a different pattern—added N increased the decomposition rate of control plot litter slightly but significantly in the 0.3 ky site, as did N + P additions at the 4100 ky site; other treatments had no significant effect (fig. 5.6e,f) (Hobbie and Vitousek 2000).

Combining the influence of litter and site quality, litter produced and decomposed in P-fertilized plots of the P-limited 4100 ky site decomposed significantly more rapidly than control litter in control plots, but there was no parallel effect of N fertilization in the N-limited 0.3 ky site—and in both sites, additions of N plus P had more substantial effects than did either nutrient alone (fig. 5.6a,b) (Hobbie and Vitousek 2000).

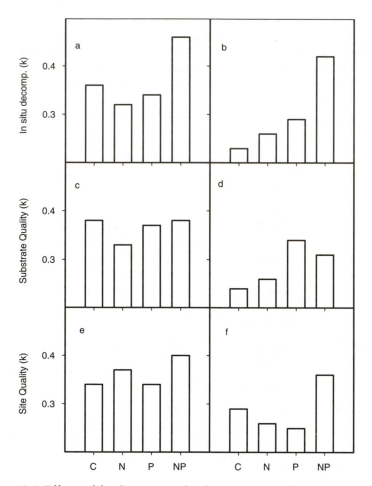

FIGURE 5.6. Effects of fertilization on the decomposition of *Metrosideros poly-morpha* leaf litter in the young and old sites, modified from Hobbie and Vitousek (2000). (a, b) Decompositon rate constants for litter collected in control and fertilized plots of each site, and decomposed in the site and treatment where it was collected. (c, d) Rate constants for litter collected in the fertilized and control plots of each site, and decomposed in control plots as a measure of litter quality or decomposability. (e, f) Litter collected in the control plots of each site, and decomposed in control and fertilized sites as a measure of site effects on decomposition. N additions to the N-limited young site have small and inconsistent effects on decomposition, but P additions to the P-limited old site enhance decomposition in situ substantially.

A parallel study demonstrated that neither N nor P stimulated root decomposition in the 0.3 ky site, but that each did so independently in the 4100 ky site—P more so than N (Ostertag and Hobbie 1999). In any case, rates of litter decomposition in fertilized plots of the infertile sites never approached those observed in naturally more fertile sites (fig. 4.14); in this sense, long-term fertilization does not make forests on infertile sites function like those on fertile sites, particularly where N is the main limiting nutrient.

Nutrient Regeneration

Does long-term fertilization lead to a more rapid regeneration of nutrients into biologically available forms in soil? We found that litter produced and decomposed in P-fertilized plots of the P-limited site released a much larger fraction of its (large) initial P content more rapidly than did control litter (fig. 5.7d); much of this release occurred early in decomposition, probably via leaching (Hobbie and Vitousek 2000). In contrast, N fertilization of the N-limited site did not cause more rapid release of N from decomposing litter, or indeed any net release of N during the first two years of decomposition (fig. 5.7a).

CONTROLS OF PLANT-SOIL-MICROBIAL FEEDBACK

Overall, fertilization with P caused the P-limited 4100 ky site to function more like forests on naturally fertile sites, at least with regard to P cycling (fig. 5.8). Increased P availability led to decreased P residence times in the forest canopy (table 5.5), decreased overall P-use efficiency (fig. 5.5), increased P concentrations in plant tissues (table 5.4) and in leaf litter (table 5.6), increased rates of litter decomposition (fig. 5.6), and more rapid regeneration of P from decomposing litter (fig. 5.7). All of these changes were statistically significant (Vitousek 1998, Hobbie and Vitousek 2000, Harrington et al. 2001); for all except the last two, the change induced by fertilization were at least as large as the difference between control plots in the 4100 ky site and naturally more fertile intermediate-aged sites.

N additions to the N-limited 0.3 ky site had quite different effects (fig. 5.8). Although long-term N fertilization decreased N residence time in the forest canopy (table 5.5), overall N-use efficiency increased slightly (fig. 5.5). N concentrations in plant tissues (table 5.4) increased significantly, but never approached those in naturally more fertile intermediate-aged sites. Neither decomposition in situ (fig. 5.6) nor rates of N regeneration from decomposing litter (fig. 5.7) responded significantly to long-term N fertilization. Accordingly, N fertilization did not cause the forest on the N-limited 0.3 ky site to function like forests on naturally more fertile sites.

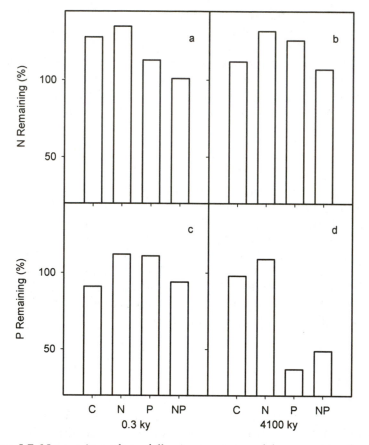

FIGURE 5.7. Net nutrient release following two years of decomposition for *Metrosideros* litter collected and decomposed in control and fertilized plots, from Sarah Hobbie (personal communication). (a, b) N remaining within litter in the young and old sites; N was immobilized in most treatments, and there was no net N release anywhere. (c, d) P remaining in the young and old sites; within two years, most of the P originally present in litter had been released back into circulation in the P-fertilized plots of the P-limited 4100 ky site.

While a positive feedback clearly contributes to patterns of N as well as P cycling along the age gradient (chapter 4), most of the components that contribute to this feedback do not respond to additions of N. Why not? What constrains responses to N additions, such that tissue chemistry and rates of decomposition/nutrient regeneration never match those in naturally fertile intermediate-aged sites, even after more than a decade of N fertilization?

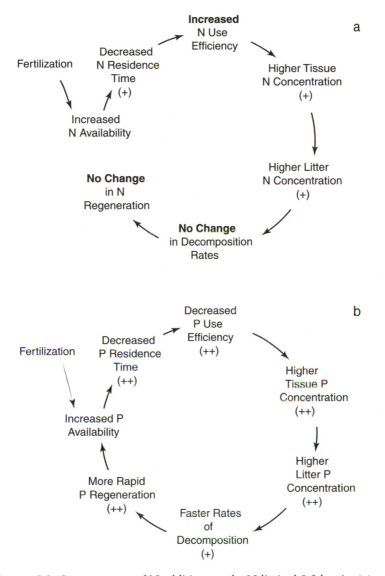

FIGURE 5.8. Consequences of N additions to the N-limited 0.3 ky site (a), and of P additions to the P-limited 4100 ky site (b), for components of the plant-soil-microbial feedback outlined in fig. 4.17. Where fertilization changed a component in the expected direction, the symbols in parentheses represent the strength of the effect, with + representing a significant change that is not as large as the difference between infertile and fertile sites along the substrate age gradient and ++ representing responses that are at least as large as those along the gradient. Where fertilization did not change a component in the expected direction, the response that was observed is boldfaced.

I will address these questions at some length in chapter 8. Briefly, I see several possible reasons for the lack of responsiveness to added N. First, although tree growth might be limited primarily by N in the 0.3 ky site and by P in the 4100 ky site, other ecosystem processes could be limited by different nutrients (Hobbie and Vitousek 2000, Sundareshwar et al. 2003). For example, decomposition rates in situ responded substantially more to additions of N plus P than to either alone, in both sites—and while background N availability is relatively high in the 4100 ky site (fig. 4.5), background P availability in the 0.3 ky site is relatively low (fig. 4.4b). Second, Kathleen Treseder demonstrated that genetically-distinct populations of *Metrosideros* occupy different sites on the age gradient (Treseder and Vitousek 2001b). Perhaps N concentrations in plant tissues remain low following fertilization because the populations that occupy infertile sites cannot accumulate as much N as those on fertile sites, no matter how much is available. If so, continuing to add nutrients for a sufficiently long time might lead to replacement of the dominant population of those sites, presumably by one that is more responsive to added N (Stemmermann 1983). Alternatively, other N-demanding species might invade the enriched site (Chapin et al. 1986, Ostertag and Verville 2002). Either way, development of a plant-soil-microbial feedback following fertilization would require both increased N availability and colonization by organisms that use N differently from the populations that now occupy the site.

Third, the decomposition of litter produced on infertile sites generally is thought to be controlled more by the abundance of recalcitrant C compounds (e.g., lignin and polyphenols) than by nutrients (Paul and Clark 1996). Consistent with that generalization, Hobbie (2000) demonstrated that while N addition enhances the decomposition of low-lignin *Metrosideros* litter substantially, it has little effect on the decomposition of higher-lignin litter like that produced along the age gradient (fig. 5.9). Studies elsewhere also have reported wide variation in the responsiveness of litter decomposition to N additions, even at sites where N limits NPP (Fog 1988, Hunt et al. 1988, Van Vuuren and van der Eerden 1992, Prescott 1995). In Hawai'i, adding P to P-limited forests reduces concentrations of lignin and polyphenols in leaf litter, but N additions do not affect recalcitrant C compounds in litter of N-limited sites (table 5.6) (Hättenschwiler et al., 2003).

Fourth, a consistent litter type decomposed more rapidly when it was placed in a fertile intermediate-aged site than in an infertile young or old site (fig. 4.16), but it responded much less (if at all) when placed in N and P fertilized plots (fig. 5.6e,f). Why? Perhaps this difference could reflect the role of different decomposer communities in fertile versus infertile sites. Several studies have observed differences in soil microbial and faunal communities across the age gradient (Nusslein and Tiedje 1998, Hobbie and Vitousek 2000, Balser 2002, R. Bardgett, personal communication,

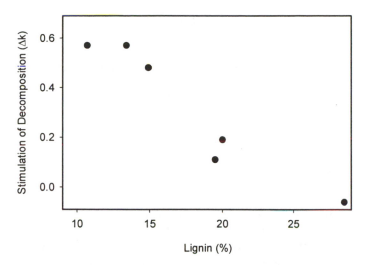

FIGURE 5.9. The relationship between litter lignin concentration and the effect of added N on decomposition, from information in Hobbie (2000) and Hobbie and Vitousek (2000). Decomposition of high quality (low lignin) litter is stimulated by added N, but decomposition of low quality (high lignin) litter is not affected by added N.

D. Foote, personal communication), but it has been difficult to map these differences in composition onto differences in function—in Hawai'i as elsewhere.

Any or all of these mechanisms might constrain the responsiveness of N cycling to N additions—but most of them leave open the question "why is the cycling of P within these ecosystems more responsive to P additions than is the cycling of N to N additions?" Biochemically, it might be easier to store additional P as phosphate than it is to store N as amino acids or other C-containing compounds, and so easier to build up tissue concentrations of P. Also, added N could allow plants to gain more C and so maintain high lignin/polyphenol concentrations, while added P could allow cell division and growth, diluting the concentrations of recalcitrant C compounds. Both N and P are required for both photosynthesis and growth, of course, but N-limited plants could be relatively more constrained in their ability to gain C, while P-limited plants could be relatively more constrained in cell division.

Alternatively, P limitation could be a part of the evolutionary heritage of *Metrosideros*, in a way that N limitation is not. *Metrosideros* originated in the southwest Pacific (Wright et al. 2000, 2001), and many of its relatives outside Hawaii occupy old, deeply leached, P-depleted soils. Perhaps the relative responsiveness of *Metrosideros* to P reflects that heritage; perhaps a lineage with its past in N-limited boreal systems (e.g., *Betula*) would be more responsive to N than to P.

Chapter Six

NUTRIENT INPUTS TO HAWAIIAN

ECOSYSTEMS: PATHWAYS, RATES,

AND CONTROLS

ON THE TIME SCALE of months to years, the total quantity of most nutrients in terrestrial ecosystems is close to constant (barring intense disturbance), and the rate at which nutrients cycle between plants, microorganisms, organic matter, and inorganic forms determines their availability to organisms. However, terrestrial ecosystems are open systems, with inputs and losses of nutrients—and in the longer term (centuries to millenia or more), the balance between nutrient inputs and outputs controls the quantity of nutrients present within the system, and ultimately their availability. Element inputs thus are essential to maintaining the fertility of soils and productivity of ecosystems over millions of years.

In this chapter, I describe pathways and rates of element inputs to each site on the Hawaiian age gradient. We anticipated that element inputs would change dramatically in their dominant pathways, rates, and ratios over the course of soil and ecosystem development (Chadwick et al. 1999). Because most pathways of input (except biological N fixation) are controlled at least partly geochemically, by processes occurring outside ecosystems, we further anticipated that inputs of particular elements—and especially ratios of elements—would bear little relation to the requirements of organisms within ecosystems. The research summarized here was motivated in part by the opportunity to understand how the geochemical stoichiometries of different pathways of element input relate to the biological stoichiometries of organisms and of nutrient cycling, in a model system that makes the analysis of both geochemical and biological processes relatively "convenient" (Krogh 1929 in chapter 2).

Inputs of Elements

The major pathways of element inputs to terrestrial ecosystems include the chemical weathering (partial or complete dissolution) of minerals in rocks and soil, deposition of elements from the atmosphere, biological fixation of N_2, and in particular circumstances the transport of elements from upslope, and transport by animals (Likens et al. 1977). Defining the major pathways for element inputs is straightforward, but measuring them quantitatively is far more difficult, and understanding their controls is a major challenge. Fortunately, several features of Hawaiian ecosystems in general, and of the age gradient in particular, make it possible to measure rates of element inputs and/or to determine time-integrated sources of elements more straightforwardly than is possible in most ecosystems.

WEATHERING

When rock arrives at the surface of Earth, it contains an assemblage of minerals that generally formed under conditions of higher temperature and pressure. At the surface, these original minerals (termed primary minerals) are exposed to fluxes of liquid water and acidity—and they weather chemically, dissolving partially or completely (congruently). The breakdown of primary minerals liberates some elements into soluble forms that are accessible to organisms, and that also can be lost to streamwater or groundwater. Some of the soluble products of weathering combine with others, or with the incompletely dissolved residue of primary minerals, to form secondary minerals within soils; most soil clays are secondary minerals. Over time, secondary minerals themselves can weather, breaking down once other sinks for soil acidity are exhausted. Ultimately, only a residue of insoluble and close-to-inert material remains. Weathering is thus a multi-step process, with the initial breakdown of readily weathered primary minerals followed by successive dissolution and mobilization of ever more unreactive primary and secondary minerals.

Immediately after a volcanic eruption in Hawai'i, the dominant minerals in the lava substrate are glass, olivine, clinopyroxene, feldspar, and magnetite-ilmenite. These primary minerals weather rapidly in the rain forest environment, particularly after plants become established (Cochrane and Berner 1997). Olivine and glass are quickly consumed by congruent weathering processes, and even feldspar has nearly disappeared by the 20 ky site. As outlined in Fig. 3.10, non-crystalline minerals then form a highly reactive pedogenic mineral assemblage that persists for more than 1000 ky. Relatively unreactive kaolinite and sesquioxide clays that are typical of highly weathered tropical soils accumulate more slowly, finally dominating the soil of the 4100 ky site.

Concepts and Definitions

In this book, I treat weathering as an input of elements to ecosystems. Many studies consider weathering to be an output—and indeed, weathering is calculated based on measured element losses, both in watershed studies (Likens et al. 1977) and in soil-based studies like this one (Chadwick et al. 1990, Brimhall et al. 1992). I prefer to treat weathering as an input for two reasons. First, the elements in primary minerals are not available to organisms or to other biogeochemical processes until those minerals are weathered, and I consider an irreversible transformation from abiotic and immobile forms into available ones to represent an input of elements to ecosystems. Second, one of the virtues of the Hawaiian age sequence is that we can calculate weathering independently of current fluxes of elements out of these systems. In chapter 7, I report losses of elements by hydrologic pathways as outputs. Whatever the semantics of weathering, the release of elements from inert to dynamic forms clearly belongs on the opposite site of the ledger from these hydrologic losses.

The multi-stage nature of weathering also poses conceptual issues, in that inputs of elements via weathering can be defined as occurring either when primary minerals that contain the element of interest break down, or as occurring when both primary or secondary minerals are weathered— effectively, when the element is lost from the soil. The former approach is straightforward conceptually, but difficult to apply in practice to the dynamics of relatively immobile elements such as P. Accordingly, I define weathering as the release of elements from primary and secondary minerals.

Approaches

Because both the volume and the mass of soil can either increase or decrease as soils develop, the weathering and loss of elements cannot be determined simply from changes in concentrations of elements in soil. Hydration and added organic matter expand (dilate) soil and add mass that dilutes element concentrations. Weathering and leaching remove not only a particular element of interest, but ultimately most of the material that was present in parent material. This loss can cause collapse of the soil, and alter the denominator in any calculation of element concentrations through soil development (Brimhall et al. 1992).

These changes can be accounted for by calculating the loss of a given element relative to the quantity of an immobile index element in soil (Chadwick et al. 1990, Brimhall et al. 1992). In practice, it is difficult or impossible to demonstrate that any element is absolutely immobile, and losses of other elements are calculated relative to the least mobile element. Initially, Oliver Chadwick made a complete set of weathering calculations for Hawaiian gradients, using zirconium (Zr) as the immobile element (Chadwick et al. 1994, Vitousek et al. 1997b, Kelly et al. 1998, Chadwick et al. 1999). Later, Kurtz et al. (2000) demonstrated a degree of Zr mobility in older soils. Niobium (Nb) and tantalum (Ta) now appear to be the least mobile elements of the many that have been evaluated, and Nb is used as the basis for weathering calculations here.

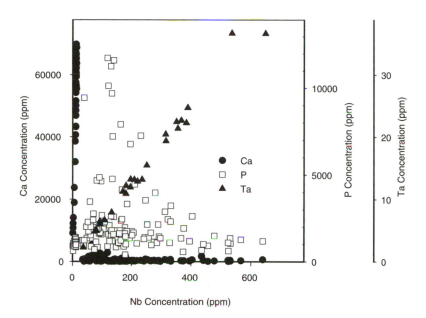

FIGURE 6.1. Concentrations of Ca (a mobile element), P (low mobility), and Ta (immobile) relative to those of the immobile index element Nb in soil horizons across the Hawaiian substrate age gradient, from information in Kurtz et al. (2000) and from Oliver Chadwick (personal communication). Increasing concentrations of Nb reflect weathering and loss of the soil matrix. Ca is lost very rapidly as weathering begins, but P concentrations increase in moderately weathered soil horizons as more mobile elements in the matrix are lost. With more intense weathering, P itself is lost. The highest Nb concentrations reflect loss of > 90% of the mass of parent material.

Fig. 6.1 shows changes in the concentration of a highly mobile element (Ca), a less-mobile element (P), and another immobile element (Ta) relative to those of Nb across the range of sites and soil horizons on the age sequence. Increasing Nb concentrations along the x-axis indicate loss of mass from the soil as a whole, with residual enrichment of Nb. Almost all the Ca is lost from soil almost as soon as any mass is lost (fig. 6.1). However, P concentrations increase in moderately weathered situations due to the loss of more mobile constituents such as Ca, and then decline with more intense weathering as P too is lost.

To estimate rates of weathering for each element, we calculate the inventory (I_{Nb}) of Nb in a soil profile:

$$I_{Nb} = \sum_{i=0 \to d} p_i \, z_i C_{Nb,i}$$

$$(6.1)$$

where d is the overall depth of the profile, p_i is the bulk density of horizon i, z is its thickness, and $C_{Nb,i}$ is the concentration of Nb in horizon i.

We calculate the inventory of an element of interest (I_j) similarly. Knowing the ratio of that element to Nb in parent material ($C_{j,p}/C_{Nb,p}$), we can then calculate the quantity of element j (Q_j) weathered and lost (relative to Nb) as:

$$Q_j = I_{Nb}(C_{j,p}/C_{Nb,p}) - I_j \qquad (6.2)$$

and the fraction of element j in parent material that has been lost (W_j) as:

$$W_j = 1 - I_j/(I_{Nb}(C_{j,p}/C_{Nb,p})) \qquad (6.3)$$

Q_j and W_j are averaged across the four or five soil profiles sampled in each site, and element ratios in parent material calculated based on the average element ratios in shield-building (for the youngest site) and post-shield Hawaiian basalts.

The inventory of Nb in soils increases monotonically with substrate age across the gradient (fig. 6.2), reflecting incorporation of unweathered

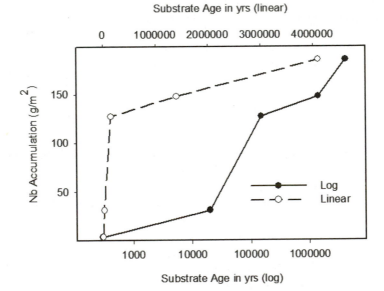

FIGURE 6.2. Accumulation of the immobile element Nb with increasing substrate age across the Hawaiian age gradient, calculated from Kurtz et al. (2001) and Oliver Chadwick (personal communication). Although most ecosystem properties along the gradient are best viewed on a log scale (solid line), the use of a linear scale here (dashed line) emphasizes that the rate of incorporation of basaltic parent material into soil declines in the older sites.

material into soil profiles at their base at the same time as upper portions of soil profiles collapse. Incorporation and collapse keep pace with each other through much of soil development, so that the depth to little-weathered material remains consistent at about one meter from the 2.1 ky site to the 1400 ky site. The weathered soil profile is much deeper (> 5 m) in the 4100 ky site, and our calculations underestimate weathering at depth there.

Finally, inputs of elements via weathering (R_j) are calculated for the age intervals between sites on the age gradient as

$$R_j = ((I_{Nb,s} - I_{Nb,s-1})(C_{j,p} / C_{Nb,p}) + (I_{j,s-1} - I_{j,s}))/(A_s - A_{s-1}) \quad (6.4)$$

where the subscript s represents a site on the age gradient, $s-1$ the next younger site, A_s the substrate age in site s, and A_{s-1} substrate age in site $s-1$. All of our weathering calculations exclude the 2.1 ky 'Ōla'a site, because much of the soil profile in that site consists of the same Keanakāko'i ash deposit that dominates the youngest site on the sequence (D. Swanson, personal communication).

Element Inputs via Weathering

Elements differ in rate and pattern of inputs via weathering depending on their abundances in Hawaiian basalt and their mobility. The mobile elements Ca, Mg, K, Si, and others are lost rapidly from young sites, and so have large inputs via weathering therein. Continued inputs via weathering in older sites are supported only by the incorporation of unweathered material at the base of soil profiles, and rates of input via weathering in the older sites decline dramatically (table 6.1). In contrast, a large fraction of the less mobile elements P, Al, and Fe remains in the soil in secondary minerals, and so our calculated inputs of these elements via weathering are lower in young sites—but relatively greater in older sites as secondary minerals themselves weather. This contrast is illustrated for Ca and P in fig. 6.3, and summarized for all major elements for the intervals from 0 to 0.3 ky versus from 1400 ky to 4100 ky in table 6.1. The ~350-fold decrease in Ca supply via weathering, versus "only" a 67-fold decrease for P, raises the question—why is it that P availability, and not the availability of a more mobile element such as Ca, limits NPP in the oldest site on the Hawaiian age gradient?

TABLE 6.1

Calculated annual inputs of elements via weathering during the initial (0.3 ky) and final (1400–4100 ky) intervals between sites along the Hawaiian substrate age gradient. All values in kg ha^{-1} yr^{-1}; from the calculations described in the text, based on information from Oliver Chadwick, personal communication.

Element	0–0.3 ky	1400–4100 ky	final/initial
Ca	43	0.12	0.0028
Mg	36	0.055	0.0016
Na	7.6	0.026	0.0034
K	3.4	0.026	0.0076
Si	281	0.65	0.0023
Al	11	0.066	0.0059
P	0.51	0.0077	0.015

ATMOSPHERIC INPUTS

Background

The term "atmospheric inputs" covers a wide variety of processes. This section considers current inputs of elements via precipitation, dry deposition, and cloudwater; other processes (e.g., long-distance transport of continental dust, biological N fixation) will be evaluated later in the chapter. The ultimate sources of elements in atmospheric deposition include marine aerosol, emissions of soluble and/or reactive gases by natural and human-caused processes, and small particles suspended by wind or by combustion. Atmospheric deposition is influenced substantially by human activity in many regions (Likens et al. 1977, Hedin et al. 1994); atmospheric transport of reactive N and S in particular has increased many-fold, especially in and near urban and agricultural areas (Galloway et al. 1995, Holland et al. 1999, Galloway and Cowling 2002).

Measurements of atmospheric inputs of elements near the 0.3 ky site were initiated in 1993, with the purpose of answering a specific question. Measurements of total N in soils of that site yielded 9800 kg ha^{-1} (fig. 4.3) (Crews et al. 1995), all of which must have accumulated since the eruptions that initiated the site. However, we could not find sources of biological N fixation anywhere near the 20–35 kg ha^{-1} yr^{-1} that this accumulation implies (Vitousek 1994) Although N budgets are notoriously difficult to balance, and dispersed sources of N fixation are difficult to measure, we wondered

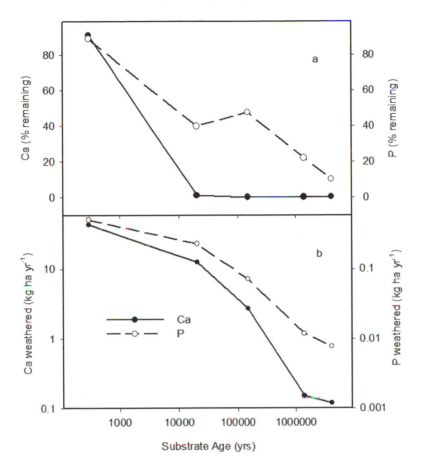

FIGURE 6.3. Weathering and loss of Ca and P across the Hawaiian age gradient, re-calculated from Chadwick et al. (1999). (a) The fraction of parent material Ca and P that remains in each site. (b) The annual input of Ca and P via basalt weathering in each site. The loss of P lags behind that of Ca due to the incorporation of P into secondary minerals, which themselves weather later in ecosystem development.

if there might be a substantial source of fixed N in atmospheric deposition. Barry Huebert and Jackie Heath Carrillo, later joined by others, addressed this question by measuring inputs of N via precipitation, dry deposition, and cloudwater. Although their initial focus was on N, they also analyzed inputs of other elements. Recently, a second deposition station was established near the 4100 ky site on Kaua'i (Carrillo et al. 2002).

Deposition Measurements

The main site for deposition measurements is a 3000 m² clearing located 1.5 km from the 0.3 ky site. The clearing is surrounded by native *Metrosideros* forest, away from paved roads; it contains a 14 m tower with line power. Heath (2001) described methods for measuring components of atmospheric deposition. Briefly:

PRECIPITATION

The volume of rainfall received is determined using a tipping bucket rain gauge; samples for precipitation chemistry are collected using an Aerochem Metrics wet/dry collector that opens only during precipitation events.

DRY DEPOSITION

Dry deposition consists of particles that sediment out of the atmosphere, and gases (e.g., HNO_3, NH_3, SO_2) that deposit on surfaces. Heath (2001) measured dry deposition using filter packs mounted under a rain shield, making collections only when relative humidity was below 95%. The HNO_3 deposition velocity was calculated using the method of Hicks et al. (1985); particulate fluxes were bounded using deposition velocities of 0.1 and 1.0 cm/s (Duce et al. 1991, Zhang et al. 2001).

CLOUDWATER DEPOSITION

Cloudwater deposition (also called "occult deposition" or "horizontal precipitation") is the most difficult component to determine—and in montane forests, where clouds often envelop the canopy, it can also be the most important (Asbury et al. 1994, Cavalier et al. 1997, Weathers and Likens 1997, Clark et al. 1998). Determining nutrient inputs via cloudwater requires measuring both cloudwater chemistry and the volume of cloudwater deposited. Cloudwater for chemical analysis was obtained with an active Teflon string droplet collector (Daube et al. 1987) mounted on the tower, while cloudwater volume was determined using a canopy mass balance approach (Juvik and Nullet 1993) within the 0.3 ky site. Carrillo et al. (2002) measured water fluxes below the canopy via throughfall and stemflow, and calculated cloudwater input (CW) as:

$$CW = TF + SF + CS + E - R \qquad (6.5)$$

where TF is throughfall, SF is stemflow, CS canopy storage (the amount of water required to saturate the canopy), E is the water that deposits on the canopy and then evaporates (during a cloud event), and R is rainfall. These components of the water budget were calculated on an event basis and then summed annually.

Inputs of Water

Over the eight years of precipitation measurements near the 0.3 ky site, inputs of rain averaged 2730 mm/yr—close to the long-term average of ~2500 mm derived from the precipitation map in Giambelluca et al. (1986). Cloudwater inputs are less certain, and determining them requires a number of calculations, corrections, and assumptions. Carrillo et al. (2002) calculated mean annual cloudwater inputs of 1640 mm (+590/–490) in the 0.3 ky site. Although Carrillo et al. justify that number clearly, both my intuition concerning the magnitude of deposition and the results of mass-balance analyses (discussed below) lead me to believe that it is too high. I will use cloudwater inputs of 1200 mm/yr here, near the lower limit of uncertainty from Carrillo et al. (2002). Half of this cloudwater evaporates from the canopy, representing a net deposition of elements but not water.

Nitrogen Inputs

Atmospheric inputs of fixed N are summarized in table 6.2 (from Heath and Huebert 1999, Coeppicus 1999, and Carrillo et al. 2002). Dry deposition is a relatively small source of fixed N, in part because it is so seldom dry (Heath and Huebert 1999). Inorganic N in precipitation averaged 0.86 kg ha^{-1} yr^{-1} over a three-year record (1998–2000), a rate comparable to estimates derived from shorter sampling intervals in nearby sites (Harding and Miller 1982, Vitousek and Walker 1989). Cloudwater is a more important—and more variable—source of N. Most cloudwater

TABLE 6.2

Atmospheric deposition of N to the 0.3 ky Thurston site, from Heath (2001) and Carrillo et al. (2002). Total Inorganic N (TIN) is calculated as NH_4-N plus NO_3-N. Values in kg ha^{-1} yr^{-1}.

	NH_4-N	NO_3-N	TIN	Org$-$N	Total N
Precipitation	0.43	0.43	0.86	0.15	1.01
Dry deposition	—	—	0.4	—	0.4
Background cloudwater[1]	2.9	2.1	5.0	0.8	5.8
Volcanic cloud water	—	—	—	—	2.4

Note

[1]Cloudwater chemistry from Carrillo et al. (2002); cloudwater volume calculated as described in the text.

events had moderate concentrations of inorganic N, but a few had concentrations as high as 17 mg l^{-1} NO$_3$–N. All of the high-concentration events were influenced by volcanic haze (vog) from an ongoing eruption of Kīlauea, a source of fixed N that is discussed in the next section. Heath (2001) assumed that cloudwater N deposition rates calculated from the more frequent, lower-concentration events represent non-volcanic sources; by my calculations, these total 5.0 kg ha^{-1} yr^{-1} of inorganic N, a rate that is applied to all of the sites. These inputs of non-volcanic N are a little higher than those obtained in other remote, foggy locations (Weathers and Likens 1997); perhaps even the "background" deposition is affected by volcanic sources (Carrillo et al. 2002). In addition to inorganic N, the limited information available suggests that organic N contributes about 0.15 kg ha^{-1} yr^{-1} of N in precipitation, and about 0.8 kg ha^{-1} yr^{-1} in cloudwater (Carrillo et al. 2002; table 6.2).

Influence of an Active Volcano

The observation of high cloudwater N inputs associated with episodes of volcanic haze (vog) surprised us. Vog is nasty stuff, with high concentrations of SO$_2$ and H$_2$SO$_4$, some HF, and (when lava is flowing into seawater) HCl and Cl$_2$—but it is not obvious why it should contain much fixed N. Flowing lava is hot enough (1100°C) to fix N$_2$ to NO thermally—but the kinetics of fixation should be slow compared to the residence time of air near hot rock (Huebert et al. 1999).

Nevertheless, the strength of the circumstantial link between vog and elevated cloudwater NO$_3$ led Huebert et al. (1999) to measure concentrations of the reactive gas NO near and downwind of active lava flows. They found concentrations in excess of 200 ppb, orders of magnitude above background levels, and concluded that thermal fixation of N$_2$ by lava is rapid and substantial, and that it must be catalyzed by components of the magma. Although NO itself is not readily deposited on vegetation, over hours to days it oxidizes to NO$_2$ and HNO$_3$ in the atmosphere, and those chemical species deposit readily. Huebert et al. (1999) found that episodes of high cloudwater N deposition occurred when the air over the island was stagnant, or when wind reversals brought the volcanic plume back over the island.

From information in Heath (2001), I calculate that the few but highly concentrated volcanically influenced cloudwater events contributed 2.4 kg N ha^{-1} yr^{-1}; I assume that a third of precipitation inputs and half of N inputs via dry deposition also represent volcanic N. This volcanic N is limited both spatially and temporally; the near-continuous activity of Kīlauea Volcano from 1983 to the present is unusual. I assume that volcanic inputs reach only the two youngest sites on the age gradient—although of course N limits primary production in those sites (fig. 5.1).

Inputs of Other Elements

Atmospheric inputs of Cl, S, Ca, Mg, Na, and K for 1998–2000 are summarized in table 6.3, which includes three years of results (1998–2000) (Carrillo et al. 2002). As for N, cloudwater is the most significant source of elements, supplying ~85% of inputs. Measurements of P, Al, Fe, and F deposition were initiated more recently; F provides an excellent tracer of volcanic influence on air masses (Benitez-Nelson et al. 2003). Precipitation and cloudwater added ~0.12 kg ha^{-1} yr^{-1} of P to the young site in 1999, with most of that derived from local volcanic sources (table 6.3; C. Benitez-Nelson personal communication). Experimental studies with a high-temperature furnace demonstrated that a small fraction of basalt P is emitted as P_4O_{10}, and could deposit on downwind ecosystems (Yamagata et al. 1991); where lava flows reach the ocean, lava-seawater interactions can cause further fluxes of P (Sansone et al. 2001).

The provenance of elements in atmospheric deposition can be partitioned into sea-salt and non-sea-salt (NSS) sources because the consistent ratios of elements in seawater (and so in marine aerosol) provide a set of expected values for their relative abundance in precipitation and cloudwater (Gorham 1961). These ratios often are referenced to Cl, but because the active volcano can generate gaseous HCl and Cl_2, I use Na—the most abundant cation in seawater—as a reference here. I assume that all of the Na in precipitation and cloudwater is from seawater, then calculate the quantity of other elements derived from the ocean using their ratio to Na. That sea-salt-derived input is then subtracted from total deposition to calculate non-sea-salt (NSS) inputs of elements (table 6.3). The very high NSS inputs of SO_4 reflect volcanic outgassing of SO_2; among cations, most Mg is marine-derived, but much of the K and Ca has an NSS source. Carrillo et al. (2002) demonstrated that much of this K input is episodic and associated with biomass burning ignited by lava flows; possible sources of the NSS Ca are discussed below.

N and P are nearly absent from surface seawater, due to strong biological demand for these elements, and the deposition of apparently non-volcanic N is far above sea-salt values. I don't know the source of this N; there are multiple volatile and reactive N-containing trace gases, and so terrestrial, marine, and anthropogenic processes all could contribute to the observed inputs. Moreover, some of this N may represent a chronic, regional influence of volcanically fixed N (Carrillo et al. 2002).

TABLE 6.3

Atmospheric deposition of major elements other than N to the 0.3 ky site. NSS (non-sea-salt) inputs represent element inputs in excess of their ratio to Na in seawater; for Cl, SO_4–S, and P, these are believed to represent volcanic sources. Carrillo et al. (2002) concluded that NSS K represents inputs from lava-associated biomass burning; the NSS Ca apparently represents a non-volcanic source, probably a soluble fraction of continental dust. Information modified from Heath (2001) and Carrillo et al. (2002) as described in the text, except for P from Benitez-Nelson, personal communication.

	precipitation kg ha⁻¹ yr⁻¹	cloudwater kg ha⁻¹ yr⁻¹	element ratios to Na (seawater)	NSS inputs in ppt	NSS cloudwater kg ha⁻¹ yr⁻¹	volcanic source kg ha⁻¹ yr⁻¹
Cl	32	190	1.8	3.3	17	20.3
SO_4–S	9.7	66.7	0.083	8.4	58.7	67.1
Ca	1.3	8.2	0.039	0.7	4.4	—
Mg	2.1	11.7	0.12	0.2	0.2	—
Na	16	97	1	—	—	—
K	1.0	5.9	0.037	0.4	2.3	2.7
P	0.045	0.075	0.00000023	0.045	0.075	0.11[1]

Note
[1] I use 0.11 rather than 0.12 (the sum of NSS precipitation and cloudwater) for the volcanic source of P, because there is a small input of P via continental dust (discussed later in this chapter).

Long-Distance Dust Transport

Background

The same winds that brought many of the progenitors of Hawaiian plants and animals from Asia (Carlquist 1982, Juvik 1998) also brings dust from Asian deserts, thereby adding another source of elements to Hawaiian ecosystems. This continental dust differs substantially from other pathways of atmospheric deposition in its source and timing, and also in that minerals in the dust must weather before the elements they contain can be taken up by organisms.

The presence of Asian dust in Hawai'i has long been recognized. Certain soil horizons are rich in quartz and mica, minerals that are not present in Hawaiian basalt and do not form as it weathers (Jackson et al. 1971, Dymond et al. 1974). Spring storms now bring detectable fluxes of Asian dust to Hawai'i (Parrington et al. 1983, Leinen et al. 1994), and marine sediment cores demonstrate that dust transport was enhanced several-fold during full-glacial as opposed to the present interglacial conditions (Nakai et al. 1993, Rea 1994).

Inputs of dust are not unique to Hawai'i; dust contributes mass and elements to soils everywhere, especially when glacial-interglacial time scales are considered (Simonson 1995). However, because Hawai'i is derived from a hot-spot in Earth's mantle, and so differs systematically in a number of characteristics from continental crust (the source of the dust), we can trace the quantity and fate of dust more straightforwardly here than elsewhere. Moreover, the context provided by the age sequence allows us to evaluate the significance of elements in dust, relative to other pathways of element input.

Methods

Kurtz et al. (2001) exploited three differences between Hawaiian basalt and continental dust to determine the quantity of dust accumulated in ecosystems along the age sequence. First, quartz is a mineral tracer of dust; it weathers slowly, and so can accumulate in soils over tens to hundred of thousands of years. Second, the relative abundance of two isotopes of neodymium, ^{143}Nd and ^{144}Nd, differs between Hawaiian basalt (−8%) and continental crust (+6%); this difference was used as an isotopic tracer. Finally, the ratio of the concentration of the rare earth europium to its neighbors in the periodic table sumerium and gadolinium differs between continental crust and basalt; this difference was used as an elemental tracer (Kurtz et al. 2000, 2001).

The accumulation of dust-derived tracers was determined in soil of the 150 ky Kohala site, where the substrate age is well-defined (chapter 3).

All three tracers are in reasonably good agreement there, and Kurtz et al. (2001) used their abundance to calculate overall inputs of Asian dust through the history of the site to be about 1.25 g m^2 yr^{-1}. This input is 3–5 times greater than dust accumulation in ocean sediments near Hawai'i (Rea 1994), a difference that probably reflects greater scavenging of dust by orographic precipitation and cloudwater over land. This calculated input represents a minimum estimate, in that any removal of dust via erosion, or via weathering of quartz and loss of geochemical tracers, would cause the true inputs of dust to be underestimated.

Element Inputs

The elemental content of Asian dust closely matches average upper crust values (Chadwick et al. 1999), so by multiplying annual dust flux by element concentrations in the upper crust, I calculated time-integrated inputs of elements to the Hawaiian age-gradient sites by this pathway (table 6.4). Differences among the sites in climate history (Hotchkiss et al. 2000), in lack of exposure to dustier full glacial conditions (for young sites), or in exposure to the very different climates and dust fluxes of the Pliocene (for the 4100 ky site) would alter estimated inputs to individual sites.

TABLE 6.4

Inputs of elements in Asian dust. Values are long-term averages over the history of the 150 ky site, in kg ha^{-1} yr^{-1}, based on multiplying dust input from Kurtz et al. (2001; 1.25 g m^{-2} y^{-1}) by average upper crust elemental concentrations derived from the Geochemical Earth Reference Model (www.earthref.org/germ).

Element	Input
Cl	0.008
SO$_4$–S	0.012
Ca	0.37
Mg	0.17
Na	0.34
K	0.35
Si	3.80
Al	0.99
P	0.0085
N	0.001
Sr	0.0042
Nd	0.00032
Nb	0.00031

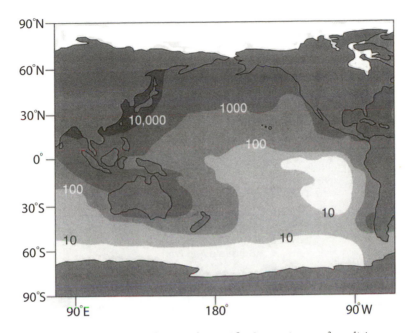

FIGURE 6.4. Inputs of Asian dust to the Pacific Ocean (mg m^{-2} yr^{-1}) integrated over glacial/interglacial cycles, reproduced with permission from Duce et al. (1991). These patterns have been validated in marine sediment cores (Rea 1994).

Overall inputs of elements via Asian dust to the Hawaiian Islands are small—among the lowest dust fluxes anywhere in the northern hemisphere. Integrated dust inputs to the western Pacific are 10–100 times greater (fig. 6.4)—and dust fluxes in continental regions are greater still (Simonson 1995). Nevertheless, inputs of most elements via dust are greater than those via basalt weathering in old sites (table 6.1), and consequently (as discussed below), elements in dust can contribute significantly to the functioning of Hawaiian ecosystems (Chadwick et al. 1999).

The Fate of Dust

Continental dust adds Nb as well as other rock-associated elements to Hawaiian soils, and these inputs could affect the calculation of basalt weathering given in table 6.1. If I assume that dust inputs have been constant and that all Nb from dust has been retained within the 4100 ky site, nearly half of the total Nb there would be dust-derived. Rates of basalt weathering would be overstated accordingly.

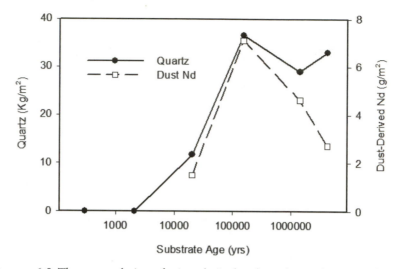

FIGURE 6.5. The accumulation of mineralogical and geochemical tracers of continental dust in soils along the Hawaiian age gradient, calculated from Kurtz et al. (2001). The lack of accumulation beyond the 150 ky site suggests loss of continental dust on time scales of < 100 ky, most likely via wind erosion (see text).

However, I believe that dust inputs have not accumulated continuously. Neither quartz nor the geochemical tracers of dust increase across the entire age gradient (Kurtz et al. 2001); rather, they increase to the 150 ky site, then remain constant or decline thereafter (fig. 6.5). This pattern could be caused by three processes: (1) dust inputs could be a relatively recent phenomenon; (2) quartz could weather and the geochemical tracers be mobilized with a residence time of ~100 ky, so that their concentrations would approach an equilibrium concentration by 150 ky; or (3) dust could be removed from the soil surface by erosion on a similar time scale (Kurtz et al. 2001). The first possibility can be ruled out by marine cores that demonstrate substantial dust inputs for the past ~3600 ky (Rea 1994). The second is unlikely: Nd in particular is relatively immobile in soils, and the 4100 ky site contains 246 g m^{-2} of Nd (Kurtz et al. 2001). If dust inputs were constant through its history and all the Nd from dust were retained, then 133 g m^{-2} of Nd should be dust-derived. In fact, only 2.7 g m^{-2} of Nd have the isotopic signature of Asian dust in this site (Kurtz et al. 2001), so most of the dust-derived Nd must have been removed over time. Leaching would not preferentially remove dust-derived Nd and leave basalt-derived Nd behind—but erosion removes the upper part of the soil where dust is deposited, and so it could have a differential effect on losses of dust versus basalt-derived Nd.

Given the geomorphology of the 4100 ky site—a remarkably flat-topped, shield-remnant ridge—I doubt that much fluvial erosion could have taken place without leaving more physical evidence. I speculate that losses of dust might have occurred via wind erosion during full-glacial times, when paleoecological studies demonstrate that Hawaiian rainforest climates were substantially drier, and fires were much more frequent (Hotchkiss 1998). In any case, dust inputs of Nb probably represent only a small (< 5%) correction to calculations of basalt weathering (fig. 6.5).

BIOLOGICAL N FIXATION

Background

Biological N fixation is the enzymatic reduction of N_2 to NH_3, which can then be used to form the many N-containing compounds that organisms require. Biological N fixation differs from other pathways of element input in that it supplies a particular element, and does so most effectively when that element is in short supply (Hartwig 1998). This capacity for biological N fixation to regulate N inputs to ecosystems makes widespread N limitation to NPP and other ecosystem processes an interesting puzzle (Vitousek and Howarth 1991, Vitousek et al. 2002)—and a topic to which I return later.

Globally, the major agents of biological N fixation on land are prokaryotes in symbioses with higher plants (especially legumes); cyanobacteria (free-living and in associations with plants and fungi); and some heterotrophic bacteria that can fix N (Sprent and Sprent 1990). The diversity of native symbiotic N-fixing plants is low in Hawai'i (Wagner et al. 1990), and none of these plants grow in the age-gradient sites, apart from a few native *Acacia koa* near the 20 ky site (Kitayama and Mueller-Dombois 1995). In contrast, many developmental sequences in continental regions include an early stage dominated by symbiotic N fixers (Stevens and Walker 1970, Reiners 1981, Chapin et al. 1994, Schlesinger et al. 1998). Elsewhere in Hawai'i, *Acacia koa* dominates some drier sites across the Islands (Pearson and Vitousek 2002), and the N-fixing lichen *Stereocaulon vulcani* is abundant on young lava flows in moist to wet environments (Kurina and Vitousek 1999, 2001).

Approach

We surveyed all of the sites for potential sources of N fixation using the acetylene reduction (AR) assay, which determines the activity of the nitrogenase enzyme (Hardy et al. 1968). Nitrogenase activity was found associated with decomposing leaf litter and dead wood, and with some bryophyte mats and lichens. Rates of N fixation in these substrates were estimated by: (1) determining the mass or cover of each substrate in each site; (2) measuring the rate of AR, repeatedly in the case of leaf litter; and (3) calibrating the AR assay by measuring fixation of ^{15}N-labelled N_2 on a subset of samples.

Rates of Fixation

Crews et al. (2000) evaluated *Metrosideros* leaf litter across the sites, and found the highest rates of AR (per g of litter) in the 2.1 ky site, followed by the 0.3 ky site; the others were substantially lower. Using a ratio of AR/^{15}N fixed of 3.9 (Vitousek and Hobbie 2000), they calculated N inputs via fixation in decomposing leaf litter of 1.2 kg ha^{-1} yr^{-1} at the 0.3 ky site, 0.6 kg ha^{-1} yr^{-1} at the 2.1 ky site (where the relatively small quantity of *Metrosideros* litter offset the high rate of AR), and < 0.2 kg ha^{-1} yr^{-1} at the four older sites.

Matzek and Vitousek (2003) surveyed the cover and AR activity of lichens, bryophyte mats, and decaying wood across LSAG; they estimated that N fixation in decomposing wood adds 0.1–0.5 kg N ha^{-1} yr^{-1}, while lichens and bryophytes together add 0.1–0.8 kg N ha^{-1} yr^{-1}. Unlike leaf litter, there is no tendency for inputs via bryophyte and lichen N fixation to be high in sites where NPP is N-limited; rather, potential N fixation appeared greatest where P concentrations in plant tissue are high and N/P ratios relatively low (Matzek and Vitousek 2003).

Overall, potential N inputs via fixation range from 0.5–1.5 kg N ha^{-1} yr^{-1}, with heterotrophic N fixers the most important sources in young sites and bryophytes and lichens more important in the 150 ky and 1400 ky sites (fig. 6.6). These rates are uncertain—particularly in the case of wood and lichens/bryophytes, which were sampled only once. Nevertheless, it seems clear that rates of N input via fixation are relatively small across the Hawaiian age gradient, substantially less than N inputs via cloudwater interception (table 6.2).

OTHER INPUTS

Additional possible sources of element inputs to Hawaiian forests include volcanic ashfalls, locally generated dust, and nesting seabirds. The magnitudes of these inputs are unknown—and the first two already are incorporated in the calculation of basalt weathering—but their possible importance should be noted. Seabirds in particular forage widely for fish and squid, and carry food back to their nests on land—and studies in other regions have shown that rates of nutrient inputs to seabird rook-

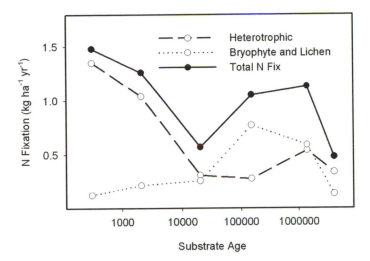

FIGURE 6.6. Inputs of N via biological N fixation across the Hawaiian age gradient, from Crews et al. (2000) and Matzek and Vitousek (2003). The upper line represents total N fixation, calculated based on measurements of acetylene reduction calibrated with $^{15}N_2$; the lower lines represent heterotrophic N fixation in leaf litter and decaying wood, and fixation associated with bryophytes and lichens.

eries can be remarkably high (Erskine et al. 1998, Anderson and Polis 1999). Prior to the arrival of humans, Hawai'i and other Pacific islands supported diverse and abundant nesting populations of seabirds on the main islands (Steadman 1995). The original Polynesian colonists collected eggs and hunted birds (Kirch 1985), and the animals that they and later colonists introduced further decimated seabird populations—effectively driving most remaining seabirds to nest on predator-free offshore islets. We don't know if seabirds were once an important source of nutrients to upland forests, but if so, their elimination by human activity could have very long-term consequences for the functioning of Hawaiian ecosystems (Loope 1998).

COMBINED INPUTS BY ALL KNOWN PATHWAYS

The pathways of element inputs described here have been characterized using different methods and integrated across very different time scales. To compare the relative importance of sources as they vary across soil and ecosystem development, I calculated inputs in terms of kg ha^{-1} yr^{-1} at each site. Inputs via basalt weathering (table 6.1) represent average rates for the time interval between two sites (eq. 6.4); because weathering declines with substrate age, these represent overestimates of current inputs at the older site. Accordingly, I estimate weathering at each site by averaging the rate from the interval before (younger than) that site with the rate from the interval after it. For the oldest site, I just used the interval before; this convention should not seriously overestimate the low rate of weathering there.

I consider inputs of elements via atmospheric deposition (precipitation, dry deposition, and cloudwater combined) to be a mixture of three components—a background flux of marine aerosol, a volcanic source of S, Cl, thermally fixed N, volatilized P, and some K from lava-caused biomass burning, and a less well-defined contribution of NSS Ca and N (Carrillo et al. 2002). The NSS Ca probably represents a soluble component of continental dust, while the N could be derived from marine, terrestrial, and some anthropogenic N trace gas emissions upwind (Carrillo et al. 2002). I want to know more about both the Ca and N sources, but for now I will treat both as part of background atmospheric deposition. I assume that the background fluxes in precipitation, cloudwater, and dry deposition are constant across all the sites, and inputs from the volcano reach only the 0.3 ky and 2.1 ky sites. Finally, I assume elements whose inputs weren't measured (Si, Sr) are deposited at the element/Na ratio of seawater.

For continental dust, input calculations are based on integrated inputs to the 150 ky site, divided by the substrate age of that site (table 6.4). I assume that these annual rates apply to all of the sites. This assumption overstates inputs to the youngest sites, which have not experienced dustier full-glacial conditions, but such inputs are very small relative to basalt weathering in young sites. Finally, I use the rates of biological N fixation summarized in fig. 6.6.

Strontium Isotopes: A Direct Test of Input Pathways

The calculated element inputs described here include substantial assumptions and uncertainties, some of which can be tested using isotopes of strontium. Sr is an alkaline earth element, and its biological and geochemical cycle is similar to those of Ca and Mg. Its isotopes ^{87}Sr and ^{86}Sr have been used to determine sources of elements in a variety of ecosystems (Graustein and Armstrong 1983, Graustein 1989, Miller et al. 1993, Stewart et al. 1998, Blum et al. 2002). Hawai'i is a particularly good place to utilize them, because the Sr isotope ratio in Hawaiian basalt is unusually homogeneous in space, time, and among minerals, and because the basalt $^{87}Sr/^{86}Sr$ ratio of 0.7036 is very different from the marine aerosol ratio of 0.7092 (Kennedy et al. 1998, Stewart et al. 2001).

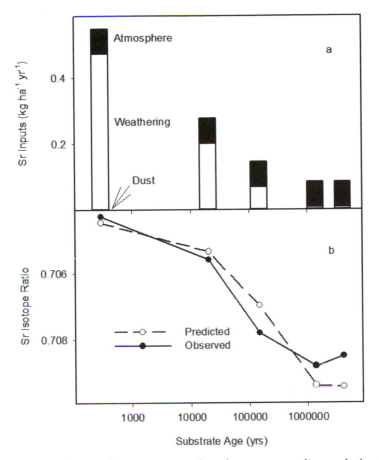

FIGURE 6.7. Pathways of Sr input across the substrate age gradient, calculated as described in the text. (a) Inputs of Sr via basalt weathering, atmospheric deposition, and continental dust (barely visible on the bottom). (b) Observed ^{87}Sr/^{86}Sr ratio in *Metrosideros* foliage (Kennedy et al. 1998), and predicted values based on weighted averages of the ^{87}Sr/^{86}Sr ratios of each input pathway. Both calculations of inputs and ^{87}Sr/^{86}Sr ratios show that basalt weathering dominates inputs to young sites; atmospheric deposition is a more important source in older sites.

Sr isotope ratios in continental dust are different but less well defined, with estimates ranging from 0.714 to 0.722 (Kennedy et al. 1998, Kurtz et al. 2000, B. Wiegand personal communication; I use 0.717 here).

I calculated annual inputs of Sr to each site along the gradient using the approach described above; results are summarized in fig. 6.7a. I then calculated the flux-weighted average of ^{87}Sr/^{86}Sr inputs by multiplying the rate of input via each pathway by its corresponding Sr isotope ratio, and dividing the result by total inputs. Because the residence time of Sr in ecosystems is short (once Sr weathers from primary minerals), I treated these flux-weighted averages as expected values for $^{87}Sr/^{86}Sr$ in biologically available Sr in each site, and compared these expected values to measured $^{87}Sr/^{86}Sr$ in *Metrosideros* leaves and exchangeable soil Sr across the age gradient (fig. 6.7b) (Kennedy et al. 1998). The match between predictions and observations is good, with Sr in the youngest site predominantly weathering-derived and Sr in the two oldest sites largely from marine aerosol. The major qualitative difference between predictions and observations is that while predictions suggest Sr in the oldest sites should be predominantly derived from marine aerosol with a small contribution from continental dust, observations reflect marine aerosol with a small contribution of basalt weathering. Overall, this analysis suggests that the assumptions underlying input calculations are reasonable, at least for Sr.

Chloride and Sulfate

Cl is the most abundant solute in sea water, and Cl inputs are dominated overwhelmingly by marine aerosol in precipitation and cloudwater (fig. 6.8a). There is a small source of volcanic Cl to the youngest sites, but thereafter I assume that all sites receive the same precipitation and cloudwater inputs of Cl. SO_4–S has a much larger volcanic source than does Cl, and also a small contribution from basalt weathering in the youngest site (fig. 6.8b). In the older sites, SO_4–S inputs are dominated by marine aerosol.

Mobile Cations

The major cations (Ca, Mg, Na, K) generally are considered to be rock-derived, and indeed basalt weathering is overwhelmingly the dominant source of the divalent cations Ca and Mg in the 0.3 ky site (fig. 6.9a, b). These elements are mobile in soils, however, and by the 150 ky site basalt weathering has declined substantially as a source of Ca and Mg (figs. 6.1, 6.9). Inputs via atmospheric deposition are much less than basalt weathering in the young site, but become the dominant source of divalent cations in the older sites—as $^{87}Sr/^{86}Sr$ ratios demonstrate (fig. 6.7b). Continental dust (as defined and measured here) always represents a very small fraction of inputs of these elements. However, as discussed above, non-sea-salt Ca contributes half of atmospheric deposition (Carrillo et al. 2002), and this "excess" Ca might represent a soluble fraction of continental dust.

In contrast, Na is extremely abundant in seawater and so in marine aerosol—and cloudwater in particular dominates Na inputs across the

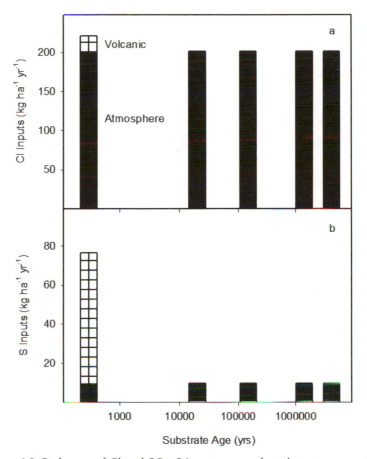

FIGURE 6.8. Pathways of Cl and SO_4–S inputs across the substrate age gradient, calculated as described in the text. (a) Most inputs of Cl are via atmospheric deposition, which I assume to be constant across the sites; there is a small input of volcanic Cl in the youngest site. (b) Volcanic sources dominate SO_4–S inputs in the youngest site; thereafter atmospheric deposition of marine aerosol dominates.

age gradient (fig. 6.10a). Basalt weathering contributes about 40% of K inputs to the 0.3 ky site (versus 90% for Ca and Mg), but atmospheric deposition is the dominant source of K thereafter (fig. 6.10b). The older sites are ombrotrophic in their cation supply, deriving most of their cations from the atmosphere (and ultimately the ocean) rather than from the underlying rock.

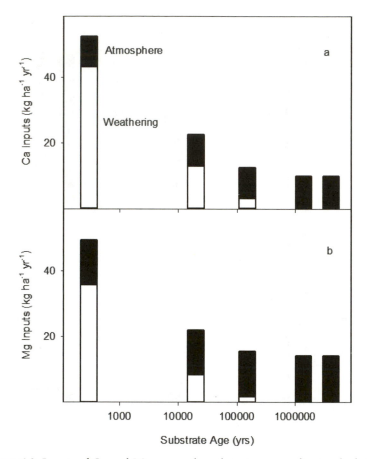

FIGURE 6.9. Inputs of Ca and Mg across the substrate age gradient, calculated as described in the text. Basalt weathering is the dominant source of Ca (a) and Mg (b) in young sites, but atmospheric deposition is the largest input to sites > 100 ky.

Silicon and Aluminum

Basalt weathering dominates Si and Al inputs in the young sites (fig. 6.11a, b). Although the weathering source of Si declines as rapidly as does that of the divalent cations, Al is retained in secondary minerals, and weathering remains a source of Al to much later in soil development. For both Si and Al, continental dust becomes an important pathway of element input once basalt weathering is exhausted (Kurtz et al. 2001).

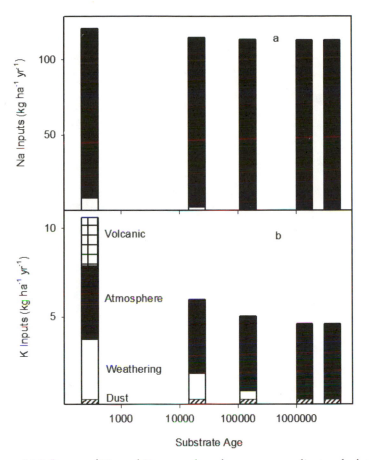

FIGURE 6.10. Inputs of Na and K across the substrate age gradient, calculated as described in the text. (a) Atmospheric deposition dominates inputs of Na, the most abundant cation in sea salt. (b) In the youngest site, volcanic sources, atmospheric deposition, basalt weathering, and even continental dust all contribute to K input, but atmospheric deposition is the dominant source of K in all of the older sites.

Nitrogen and Phosphorus

N and P are the elements that most often limit biological processes on the Hawaiian age gradient (fig. 5.1), as elsewhere. They differ strikingly in their major sources, with N being largely atmospherically derived and so renewable, and P being rock-derived and depletable (Walker and Syers

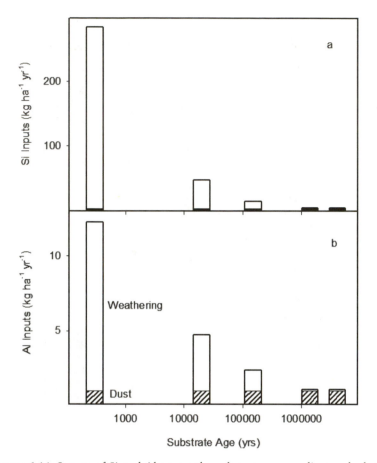

FIGURE 6.11. Inputs of Si and Al across the substrate age gradient, calculated as described in the text. Si (a) and Al (b) are derived primarily from basalt weathering in the young sites; continental dust becomes the dominant source in the oldest sites.

1976). Across the age gradient, N inputs are greatest in the youngest site, due to atmospheric deposition of volcanically fixed N and to slightly enhanced biological N fixation (fig. 6.12a). However, differences among the sites are relatively small, and cloudwater in particular represent the largest source of N (ca. 5.8 kg ha^{-1} yr^{-1}) across all of the sites. These N inputs are relatively large, compared to other remote sites, and inputs to older sites could be overstated here if the "background" N deposition in cloudwater and precipitation has been influenced by volcanically-fixed N in the 0.3 ky site (Carrillo et al. 2002)—as limited measurements of pre-

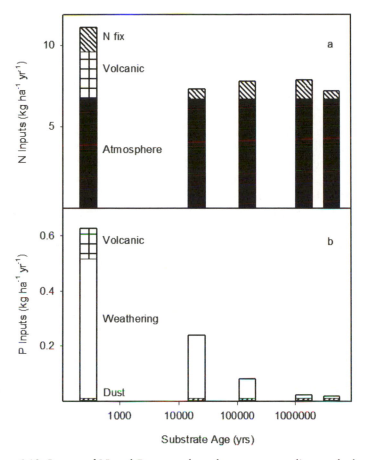

FIGURE 6.12. Inputs of N and P across the substrate age gradient, calculated as described in the text. (a) Atmospheric deposition is the dominant source of N along the gradient; biological N fixation contributes a relatively small quantity of N in all of the sites, and volcanically fixed N is a significant source in the youngest site. (b) Basalt weathering is the dominant source of P in young sites, with a small contribution from volatile volcanic sources in the youngest site, while continental dust contributes the majority of P received by the oldest site.

cipitation chemistry near the 4100 ky site suggest. Moreover, the pattern of N inputs could differ substantially if the young sites were dominated by a symbiotic N fixer, as occurs in many soil developmental sequences.

In contrast to N, inputs of P decrease by a factor of ~40 once the basalt weathering source is nearly exhausted (fig. 6.12b). The ratio of N:P in inputs drops from ~20:1 on a mass basis in the youngest site, which is close to the ratio of 10–14:1 in *Metrosideros* foliage, to ~400:1 in the oldest site. Inputs of continental dust make up a trivial fraction of total P inputs in the youngest site—but they represent the majority of P inputs to the oldest site, where P supply demonstrably limits NPP.

A comparison of P inputs (fig. 6.12b) with those of cations (fig. 6.9, 6.10) makes clear why P rather than cations limits biological processes on old substrates in Hawai'i—despite the much greater mobility and loss of the cations (fig. 6.3) (Chadwick et al. 1999). Atmospheric deposition supplies cations but not P from marine aerosol—enough cations so that when the basalt weathering source is depleted, the decrease in P inputs is proportionately much greater than that in cations. Indeed, were it not for the trickle of P that travels > 6000 km in dust from Asia, much of it during full-glacial periods, I'm not sure it would be possible to maintain forests on the oldest substrates.

The significance of continental dust to the P economy of Hawaiian forests on old soils is a striking result of the Hawaiian study (Chadwick et al. 1999), but there is no reason to think that long-distance dust transport is uniquely important here. Hawai'i is after all among the least dusty places in the northern hemisphere (fig. 6.4), and all the major continental rainforests receive more dust than does Hawai'i (Swap et al. 1992, Rea 1994, Stoorvogel et al. 1997). Dust from deserts probably contributes substantially to the long-term P budget of all ecosystems on old, deeply leached soils. Continental dust is even more important in marine systems; it represents the major source of Fe in the oligotrophic North Pacific— where Fe supply often limits biological N fixation, causing N supply to limit marine productivity (Karl 1999, Karl et al. 2002). Moreover, the much dustier conditions of full-glacial times could have stimulated open-ocean productivity substantially, contributing to the draw-down in atmospheric CO_2 recorded in full-glacial ice cores (Petit et al. 1999, Sigman and Boyle 2000).

Chapter Seven

NUTRIENT OUTPUTS: PATHWAYS,

CONTROLS, AND INPUT-OUTPUT BUDGETS

WITH THE EXCEPTION of biological N fixation, the pathways of element input described in chapter 6 deliver a suite of elements that bears little relation to the requirements of organisms. In contrast, where the supply of a particular element limits productivity or other biological processes, organisms actively take up and retain that element, and so its losses should be small (Vitousek and Reiners 1975, Rastetter and Shaver 1992). However, elements that are not required by organisms—and biologically essential elements that are abundant enough to satiate the demands of organisms—generally should have outputs close to inputs. In this sense, element outputs from terrestrial ecosystems can be regulated biologically, in a way that most pathways of element inputs cannot.

This analysis suggests that the elements lost from terrestrial ecosystems are those that organisms do not need. However, recent research demonstrates that biologically essential elements can be lost from terrestrial ecosystems in forms that are unavailable (or only slowly available) to organisms—and that such losses can contribute to maintaining nutrient limitation (Hedin et al. 1995, Vitousek et al. 1998). The best-characterized pathway of loss that fits this description is the leaching of dissolved organic N (DON). Based on research in unpolluted forests in southern Chile, Hedin et al. (1995) suggested that: (1) DON is the dominant pathway of N loss from these forests; (2) most of the DON that leaches from forests is not available to organisms; and (3) losses of DON could suffice to constrain N accumulation in ecosystems at a level that sustains N limitation. In effect, losses of N in a form that organisms cannot get keeps inputs of N from accumulating to the point that N no longer limits biological processes.

Several lines of evidence support portions of this conceptual model. Losses of DON vary much less than do losses of dissolved inorganic N (DIN) in the form of ammonium- and nitrate-N, among ecosystems and experimental treatments that differ in N availability (McDowell and Asbury 1994, Lajtha et al. 1995, Currie et al. 1996, Goodale et al. 2000, Perakis and Hedin 2002). Several studies demonstrate that high molecular weight DON is relatively recalcitrant to biological uptake and breakdown (Qualls et al. 1991, 2000; Neff et al. 2000, Perakis and Hedin

2001—but see Northrup et al. 1998). In contrast, amino acids are readily acquired and metabolized by plants as well as microbes (Chapin et al. 1993, Nasholm et al. 1998). Finally, both simple and relatively complex ecosystem models demonstrate that losses of DON could indeed constrain N accumulation in a way that sustains N limitation to NPP, at least in ecosystems where symbiotic N fixers are sparse or absent (Vitousek et al. 1998, Vitousek and Field 1999).

The Hawaiian substrate age gradient offers a range of unpolluted systems that differ widely in N availability (fig. 4.5), in which controls of DIN versus DON losses can be evaluated while keeping other factors constant. In addition, the pattern of N inputs and changes in N cycling along the gradient can be used to evaluate whether DON losses suffice to constrain N accumulation and sustain N limitation. These features motivated Lars Hedin, Pamela Matson, and me to evaluate the patterns and controls of N losses from all of the sites along the age sequence. We included fluxes of N to the atmosphere as well as leaching losses, and we tested the applicability of the inorganic/dissolved organic dichotomy to losses of P in addition to N. In the course of these analyses, we measured leaching losses of all of the major anions and cations across the age gradient (Hedin et al. 2003).

Output Pathways

The major pathways of element loss from terrestrial ecosystems are: (1) leaching of dissolved and colloidal material through soils to streams and/or groundwater; (2) for C, N, and S in particular, fluxes to the atmosphere via volatilization, denitrification, and other transformations; and (3) erosion, the transport of particulate material by gravity, wind, and water. Other pathways can be important in particular systems, including harvest, volatilization and/or suspension by fires, and the movement of animals.

Leaching

The two major techniques for measuring leaching losses of elements are small watersheds and lysimeters. The small watershed approach makes use of places where the bedrock underlying terrestrial ecosystems is watertight, and stream channels are well defined. It is based on careful measurements of both streamflow and solute concentrations, and calculates element losses via hydrologic pathways as their product (Likens et al. 1977). This approach has the great advantage that fluxes of water are measured directly, and so overall hydrologic losses are well-constrained. However, where the major purpose is determining element losses from the rooting zone of a forest, rather than from the entire forest/riparian zone/stream ecosystem, it has the disadvantage that it includes processes that occur below the reach of terrestrial biota (eg. chemical weathering deep in the soil, riparian denitrification, and P spiraling in streams). It has the further disadvantage of not being applicable in many places—including the Hawaiian Islands, where the bedrock rarely is watertight.

In contrast, lysimeter-based approaches involve placing collectors for soil solution at a defined depth within the soil, calculating water flux past that depth using a hydrologic model of some sort, and determining outputs as the product of calculated water flux and measured chemistry in the collectors. The approach has the disadvantage of being relatively unconstrained hydrologically, but the advantage that it can be done at any depth, or several depths, in the soil—and the further advantage of being feasible in Hawaiian forest ecosystems.

We installed at least six soil-water collectors below the depth of most of the roots in each site. The mean radiocarbon age of the soil at this depth was > 2000 yrs in all except the two youngest sites (Torn et al. 1997), suggesting that most biological activity occurs above this depth. These lysimeters pull water from the soil matrix under tension; to obtain samples, we visited each site approximately monthly, applied a vacuum to each lysimeter, and collected solution samples 48–72 hours later. All samples were analyzed for pH, alkalinity, Ca^{++}, Mg^{++}, K^+, Na^+, SO_4^{2-}, Cl^-, NH_4^+, NO_3^-, PO_4^{3-}, dissolved organic N (DON), and dissolved organic P (DOP); a subset of samples was analyzed for Si, Al, and dissolved organic C (DOC).

The balance between cations and anions in lysimeter solutions provides a check on the accuracy and completeness of analyses of dissolved ions. The anion/cation balance was close in all sites; we assume that the small excess of cations over measured anions (fig. 7.1) reflects the influence of soluble organic acids, which were not measured directly (Hedin et al. 2003).

We calculated the volume of water leaching through soil by estimating evapotranspiration (~800 mm/yr) and subtracting that from incoming precipitation plus cloudwater. Here, I modified the calculations in Hedin et al. (2003) by including 600 mm/yr of cloudwater inputs to each site, as described in chapter 6. Element concentrations in lysimeters did not vary seasonally, and indeed differed relatively little in time or space within a site compared to the large differences among sites. Accordingly, we estimated annual losses by multiplying calculated annual water fluxes (precipitation + cloudwater – evapotranspiration) by mean ion or element concentrations in each site.

In addition, we collected and analyzed samples from small streams near the four oldest sites on the sequence (there are no streams on young

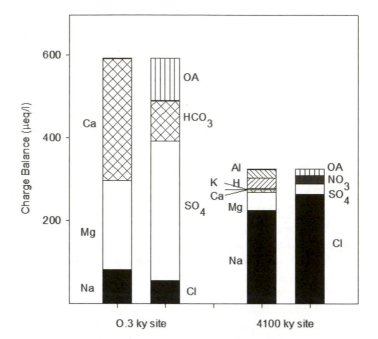

FIGURE 7.1. Cation/anion balance in lysimeter solutions of the youngest and oldest sites on the Hawaiian age gradient, revised from Hedin et al. (2003). The dominant ions in the young site are derived from volcanic activity (SO_4–S), and basalt weathering (Ca and Mg); the major ions in the old site come from marine aerosol (Na and Cl).

substrates). We used these analyses to evaluate whether lysimeters and streams sample similar components of the water passing through these ecosystems, and to estimate the importance of basalt weathering and other processes that occur below the depth of the lysimeters.

For most of the elements, leaching was the only pathway of loss that we measured. Accordingly, I will discuss the magnitude and controls of N and P losses via various forms and pathways, and then present leaching losses of most elements in the context of input-output budgets at the end of this chapter.

N-Containing Trace Gases

Several N-containing gases can be emitted from terrestrial ecosystems, including ammonia (NH_3), dinitrogen (N_2), nitrous oxide (N_2O), and nitric oxide (NO). We used a chamber-based approach to measure fluxes of N_2O–N and NO–N across the age gradient on 15–25 separate days spread

TABLE 7.1

Fluxes of N to the atmosphere from sites across the Hawaiian age gradient, in kg ha^{-1} yr^{-1}. Data from Hedin et al. (2003).

Site Age (ky)	N_2O–N	NO–N	N_2
0.3	0.01	0.01	0.27
2.1	0.01	−0.01	—
20	0.84	0.37	—
150	0.08	0.28	—
1400	0.76	0.01	4.1
4100	0.43	1.49	—

throughout a two-year period in the 0.3 ky, 20 ky, and 4100 ky sites, and four times each in the 2.1 ky, 150 ky, and 1400 ky sites. We found no seasonal variations in trace gas fluxes, and so estimated annual losses of N as the mean of the measured fluxes over all sampling times in each site (table 7.1) (Hedin et al. 2003). Ralph Riley had earlier measured fluxes in three of the sites, and found similar rates of loss (Riley and Vitousek 1995). Because N_2O is stable in the atmosphere, all of the N_2O–N flux across the soil-air interface represents N lost from the system, but an unknown proportion of the NO–N flux is absorbed within the forest canopy (Lerdau et al. 2000).

NO and N_2O can be produced as a byproduct of either nitrification or denitrification. We used acetylene to selectively inhibit these processes and so to determine the processes controlling fluxes (Riley 1996). Across the sites, most fluxes of both NO and N_2O were derived from nitrification (Hedin et al. 2003); denitrification contributed to N_2O fluxes only in the 0.3 ky and 1400 ky sites. We used the increase in N_2O fluxes in the presence of high acetylene concentrations to calculate N_2 fluxes of 0.27 kg ha^{-1} yr^{-1} and 4.1 kg ha^{-1} yr^{-1} in the 0.3 ky and 1400 ky sites, respectively. Because of sparse sampling and extremely high variability, these N_2 fluxes are included in table 7.1 but not carried through subsequent analyses of N losses (Hedin et al. 2003).

In this chapter, I focus on trace gas fluxes and their controls from the perspective of ecosystem-level N losses along the Hawaiian age gradient. However, these fluxes also have regional and global implications. N_2O is a stable, radiatively active gas that is increasing in concentration and contributing to climate change (Kroeze et al. 1999), and NO is a chemically reactive gas that drives the formation of tropospheric O_3 (Chameides et al.

1994). Their roles in the atmosphere and climate system have motivated flux measurements in many ecosystems (Matson and Goldstein 2000). Where flux measurements are grounded in an understanding of their ecosystem context and supported by process measurements—as in a number of studies in the Hawaiian Islands (Matson and Vitousek 1987, Vitousek et al. 1989, Riley and Vitousek 1995, Matson et al. 1996, Riley 1996, Hall and Matson 1999, 2003)—they can contribute substantially to understanding the sources and dynamics of globally significant trace gases.

Erosion

The age gradient sites were selected to represent stable constructional surfaces of shield volcanoes; they show little or no visible sign of erosion. No doubt some erosion has occurred, as suggested by the lack of accumulation of tracers of continental dust (fig. 6.5), but I have no basis for estimating rates of erosion—other than that they are small. Away from these stable constructional surfaces, erosion becomes a dominant process of mass removal over a progressively larger fraction of the Hawaiian landscape, as volcanoes and islands age (fig. 2.5). On older surfaces, erosion should remove nutrient-depleted surface soils, and expose deeper material to weathering (Silver et al. 1994, Scatena and Lugo 1995, Vitousek et al. in press). This removal could in effect rejuvenate the supply of rock-derived nutrients on old substrates by enhancing rates of basalt weathering within the rooting zone; it could also cause nutrient availability to be heterogeneously distributed across the landscape.

We tested the potential for erosion to rejuvenate the supply of rock-derived nutrients by measuring Sr isotope ratios in *Metrosideros* leaves along transects from stable constructional surfaces to streams in steep-sided valleys near the 150 ky and 4100 ky sites (Vitousek et al. in press, Porder et al. submitted). We found greater inputs of rock-derived nutrients on slopes; trees in lower slope positions near the 150 ky site obtained up to 85% of their Sr from basalt, versus 25% for those on adjacent constructional surfaces (fig. 7.2) (Porder et al. submitted).

Other Pathways of Loss

None of the age gradient sites appear to have been harvested or otherwise disturbed substantially by human activity, so I believe that harvest can be neglected as a substantial pathway of nutrient output. Fire may have had a greater influence; although none of the sites have burned historically (and climatically similar Hawaiian forests burn extremely rarely), fires were far more frequent around the last glacial maximum near the 150 ky site (Hotchkiss 1998), and probably in all of the older sites. I cannot rule

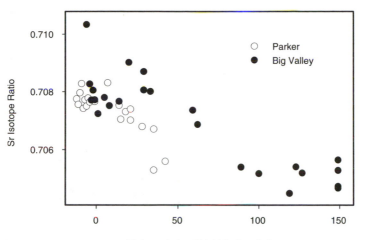

FIGURE 7.2. Sr isotope ratios in foliage of *Metrosideros polymorpha* collected along transects from shield volcanic surfaces to streams in two steep-sided stream valleys on Kohala Mountain, from Porder et al. (submitted). The Parker transect is near the 150 ky age-gradient site; the Big Valley transect is in a wetter area (3500 mm/yr). Basalt weathering releases Sr with $^{87}Sr/^{86}Sr$ of 0.7036, versus the marine aerosol ratio of 0.7092. Most Sr on and near the shield surface is atmospherically-derived, as on the age-gradient site (Kennedy et al. 1998), but basalt weathering contributes the majority of Sr inputs on the lower slopes.

out the possibility that combustion losses contribute to the element balances of these sites on long time scales.

Unlike harvest and fire, other disturbances (e.g., canopy diebacks, hurricanes) do not systematically remove elements from sites directly, but losses of elements via leaching and trace gas flux are enhanced in the years following disturbance (Likens et al. 1970, Bormann and Likens 1979). All of the older sites must have experienced multiple episodes of stand-level disturbances during their history; we know that the forest on the 4100 ky site was damaged by hurricane Iniki in 1992 (Herbert et al. 1999) and by hurricane Iwa in 1982. Although Riley and Vitousek (2000) did not find enhanced N_2O flux three months after hurricane Iniki, I suspect that cumulative losses following episodic disturbances contribute to long-term nutrient balances of these sites. Finally, even though animals probably affect rates of nutrient cycling in these sites—and introduced pigs and earthworms in particular could represent disturbances that enhance rates of element loss (Singer et al. 1984, Vitousek 1986, Bohlen et al. in press)—I doubt that either native or introduced animals have been important vectors for net losses of elements from these sites.

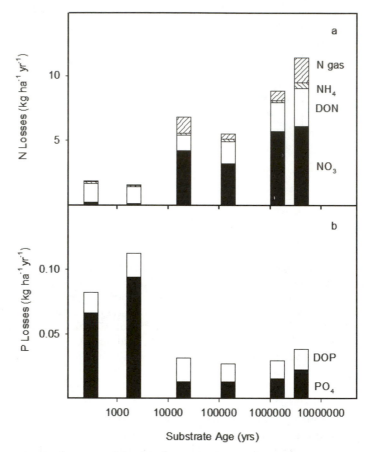

FIGURE 7.3. Pathways and forms of N and P losses along the Hawaiian age gradient, modified from Hedin et al. 2003. (a) The loss pathways for N are leaching as NO_3–N, dissolved organic N (DON), NH_4–N, and N trace gas emissions (N_2O–N plus NO–N). Leaching of DON is the major loss pathway in young sites, while leaching of NO_3–N represents the dominant N loss in the older sites. (b) Loss pathways for P are leaching of PO_4–P and dissolved organic P (DOP).

RATES AND CONTROLS OF N AND P LOSSES

P and especially N can be lost by multiple pathways—and as discussed above, differences in how these pathways are regulated could have substantial implications for nutrient cycling and limitation (Hedin et al. 1995, Vitousek et al. 1998, Hedin et al. 2003). Fluxes of N trace gases and leaching losses of NO_3–N are small in the two youngest sites, then markedly greater in the four older sites (fig. 7.3a). These losses correlate

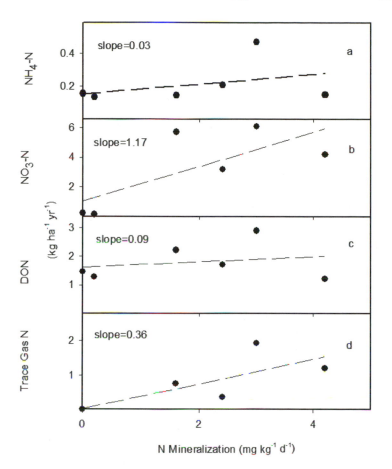

FIGURE 7.4. Correlations between rates of net N mineralization and N loss by each pathway across the substrate age gradient, modified from Hedin et al. (2003). (a) NH$_4$–N. (b) NO$_3$–N (c) Dissolved organic N (DON). (d) N trace gases. NO$_3$–N and trace gas losses correlate significantly with N mineralization and other measures of N availability.

strongly with rates of N mineralization (fig. 7.4b, d), and with the other measures of N availability discussed in chapter 4 (Riley and Vitousek 1995, Hedin et al. 2003). I conclude that where N is in short supply, biological demand for N minimizes losses of NO$_3$–N and N trace gases; losses become substantial only when N becomes relatively abundant.

In contrast, losses of DON and NH$_4$–N vary less across the gradient, and correlate poorly with N availability (figs. 7.3a, 7.4a, c). In the strongly N-limited 0.3 ky site, DON plus NH$_4$–N account for 87% of total N losses—versus an average of only 27% of the much greater losses in the four oldest sites. Although NH_4–N is available to plants and microorganisms in these sites as elsewhere (Treseder and Vitousek 2001a), perhaps the NH_4–N concentrations in leachate are so low that organisms have difficulty acquiring it. DON concentrations in lysimeters are much higher; its losses are consistent with the idea that most DON in leachate represents forms that most organisms cannot utilize rapidly (Neff et al. 2000, Perakis and Hedin 2001). Overall, losses of NO_3–N and N gases from the older sites follow the pattern I would expect for a nutrient that organisms do not need, in this case because they have enough N. The patterns of NH_4–N and DON losses are what I would expect for forms of a nutrient that organisms cannot get.

Variation in the natural abundance of ^{15}N ($\delta^{15}N$) in plants and soils across the gradient supports these conclusions. Several major pathways of loss of available N preferentially remove the lighter isotope ^{14}N, leaving residual N relatively enriched (Högberg and Johanisson 1993, Austin and Vitousek 1998, Handley et al. 1999, Martinelli et al. 1999). Sites that lose little N, or lose N only by non-fractionating pathways, tend to be ^{15}N-depleted. On the Hawaiian age gradient, soil and plant N are depleted in ^{15}N at the 0.3 ky site, where we know that losses are low and N limits NPP, but relatively ^{15}N enriched in the older sites (fig. 7.5) (Martinelli et al. 1999). Young, N-limited sites on the Mauna Loa matrix (chapter 3) also are strongly ^{15}N-depleted (Vitousek 1999).

Nutrient losses that are more or less independent of ecosystem nutrient status, as NH_4–N and DON appear to be here, have the potential to cause and/or sustain nutrient limitation (Hedin et al. 1995, Vitousek et al. 1998)—particularly where inputs of N are low. However, DON and NH_4–N losses from the Hawaiian age gradient are too small to drive N limitation. Although they dominate N losses from the young sites, total N losses are much less than N inputs there (fig. 6.12a). Accordingly, N accumulates until these systems make a transition from N-deficiency to N-sufficiency, with substantial N availability and NO_3–N loss in and after the 20 ky site. DON losses could delay the transition to N-sufficiency (Hedin et al. 2003), but they do not prevent it.

Higher rainfall forests in Hawai'i and elsewhere have relatively greater losses of DON, and could experience sustained N limitation as a consequence. A model-based analysis by Raich et al. (2000) suggests that Mauna Loa sites receiving > 4 m/yr of precipitation could be N-limited as a consequence of DON leaching, as long as biological N fixation is small or absent. Moreover, Ted Schuur (2001, Schuur and Matson 2001, Schuur et al. 2001) demonstrated that N is in shorter supply in progressively wetter sites along a moisture gradient of Maui forests receiving from 2200–5000 mm/yr; he inferred that DON losses in the wettest sites cause

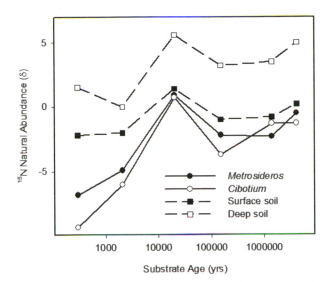

FIGURE 7.5. [15]N natural abundance in plants (solid lines) and soils (dashed lines) across the Hawaiian age gradient, from Martinelli et al. (1999). Plants and surface soils are ^{15}N depleted in the youngest sites, becoming relatively more enriched in the older sites where N availability and N losses by fractionating pathways are greater.

this pattern. Measurements along Kanehiro Kitayama's wet age gradient across Hawai'i show that those sites never achieve the higher N availability observed in intermediate-aged sites (Vitousek et al. 1995b, Kitayama et al. 1997); again DON leaching could be partly or wholly responsible.

As with N, losses of PO_4–P (dissolved inorganic P, or DIP) and dissolved organic P (DOP) follow different patterns across the age gradient (fig. 7.3b). DIP losses vary inversely with NO_3–N, being greatest in the two youngest sites, and then declining to low levels in the four older sites (Hedin et al. 2003). Losses of DIP do not parallel P availability across the gradient (fig. 4.4b); rather, they correlate with inputs of P via weathering (fig. 6.12b). This difference between N and P could reflect the low mobility of P in soils, and the fact that P availability is measured in surface soil while inputs via weathering are distributed through the soil profile in young sites.

DOP losses vary little across the age gradient, despite wide variations in DIP loss (fig. 7.3b). However, the stoichiometry of dissolved organic matter (DOM) suggests that biological processes strongly influence losses of DOP but not DON. I calculated C:N:P ratios in soil organic matter (from Crews et al. 1995), DOM in water extracts of surface soils (Neff et al. 2000), and DOM in lysimeters below the rooting zone (Hedin et al.

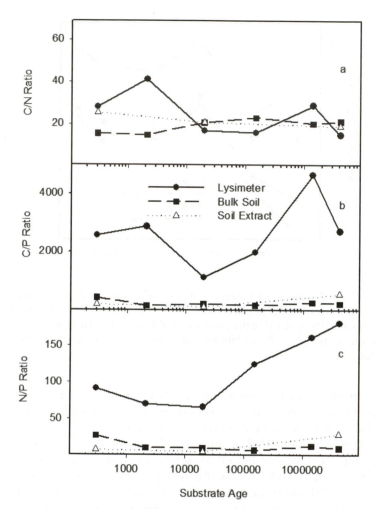

FIGURE 7.6. The C:N:P stoichiometry of soil organic matter (SOM) (Crews et al. 1995), dissolved organic matter (DOM) in extracts of surface soil (Neff et al. 2000), and DOM in lysimeters deeper in the soil profile (Hedin et al. 2003), for sites across the substrate age gradient. (a) C:N ratios do not differ systematically within or among sites. (b) C:P and (c) N:P ratios in lysimeters are much wider than those in SOM or surface-soil extracts, and wider in the low-P older sites than in nutrient-rich intermediate-aged sites, demonstrating that P is retained within these sites relative to C and N.

2003, (fig. 7.6). C:P ratios increased dramatically along this path, particularly in sites where P availability is relatively low; they reach > 2500:1 in the two oldest sites. In contrast, C:N ratios in organic matter do not change consistently along the path from SOM to lysimeters, even where N limits NPP.

I believe that this difference in the stoichiometry of DOM reflects the ability of extracellular phosphatase enzymes to remove ester-bonded P from DOM; in contrast, the removal of C-bonded N involves the coordinated activity of several enzymes (McGill and Cole 1981, Hunt et al. 1983, Olander and Vitousek 2000), in effect requiring the breakdown of DOM to release N. Nevertheless, some DOP still leaches through these ecosystems. Organisms may be unable to acquire it because it is too dilute, or has C:P ratios that are too wide, or occurs in forms that are recalcitrant to extracellular phosphatases. These losses, and losses of low concentrations of DIP that also may be too dilute for these organisms to utilize, occur even in the oldest site on the age sequence—where we know that P supply limits NPP (fig. 5.1).

INPUT-OUTPUT BUDGETS

I present input-output budgets across the gradient with considerable trepidation, because many assumptions, uncertainties, and mismatched time scales are embedded in my calculations of both inputs and outputs. Nevertheless, I believe that these rough budgets are a necessary step in moving from an understanding based on patterns and processes towards one based on integration and synthesis. The patterns in element inputs or outputs across the gradient, in this chapter and the previous one, generally can be interpreted in terms of controlling processes—and often are qualitatively similar for inputs and outputs. However, challenging this pattern-based understanding by constructing input-output budgets—thereby subjecting my analysis to the tyranny of mass balance, across millions of years of soil and ecosystem development—reveals both gaps in my knowledge and understanding of these ecosystems, and previously-unrecognized dynamics that may turn out to be important.

Calculating element budgets provides a reality check for calculations of both inputs and outputs. Without budgets, there is no independent check on the accuracy or completeness of inputs or outputs—other than predicted versus observed patterns for Sr isotopes (fig. 6.7) and the charge balance of lysimeter solutions (fig. 7.1). It is not that inputs must equal outputs; there are a number of good reasons why they might be out of balance, at many places and times. Rather, the times and places where inputs and outputs are out of balance should make ecological and geochemical sense; if they don't, then our assumptions and/or methods for measuring inputs and outputs need to be reevaluated, or our notion of ecological and geochemical sense needs to be challenged. This reality check is not widely employed, because relatively few sites permit independent estimates of the major input and output pathways.

More fundamentally, in the long term the balance between inputs and outputs of elements controls the quantity of each element in every ecosystem. This element capital in turn ultimately controls the biological availability of elements—so understanding the balance between inputs and outputs makes it possible to understand nutrient cycling and limitation in the long term.

Budget Calculations

I calculate element budgets for five of the six age gradient sites, excluding the 2.1 ky 'Ōla'a site because of the influence of the Keanakākoi ash deposit discussed in chapter 6. All the output fluxes are based on repeated measurements of lysimeter leachates and N trace gas fluxes in each site over a 2+ year period (Hedin et al. 2003). The precipitation/cloudwater inputs also were collected over a 2.5 year period—primarily in the 0.3 ky site, with some supporting information in a recently established station near the 4100 ky site (Carrillo et al. 2002). Measurements of N inputs via biological N fixation represented more of a snapshot, with three time points per site for leaf litter and one time per site for wood, lichen, and bryophyte N fixation (Crews et al. 2000, Matzek and Vitousek 2003). In contrast, inputs via basalt weathering and continental dust are based on time-integrated measurements over thousands to millions of years, as described in chapter 6. The input-output budgets calculated here thus include both short-term and integrated measurements of inputs, compared with exclusively short-term measurements of outputs.

CHLORIDE AND SULFATE

Chloride should be the most straightforward element budget to balance, especially in maritime regions. The main input of Cl is atmospheric deposition of marine aerosol (fig. 6.8a); the only significant output is leaching, and Cl is highly mobile in and through soils. Moreover, while Cl is essential to the growth of plants and animals, it is present in relatively low concentrations in plant biomass. Nevertheless, calculated Cl inputs far exceed calculated Cl outputs in the 0.3 ky and 20 ky sites (fig. 7.7a), and are close to being in balance in the three older sites. Using Carrillo et al. (2002) larger estimate for cloudwater inputs would increase calculated inputs and so widen the imbalance in young sites—and create one in the old sites.

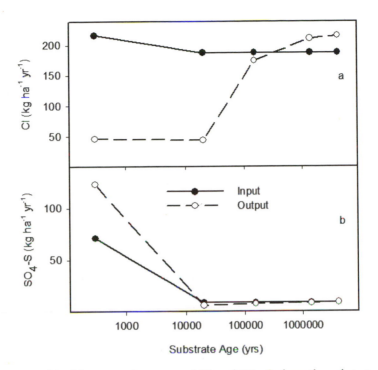

FIGURE 7.7. Total inputs and outputs of Cl and SO_4–S along the substrate age gradient. (a) Cl inputs (from fig 6.8a) are assumed to be relatively constant across the sites, but outputs (from Hedin et al. (2003)) are much lower in young than old sites. (b) SO_4–S inputs (from fig. 6.8b) are high in the youngest site, reflecting volcanic inputs; outputs are even greater than inputs in the young site, reflecting absorption of SO_2 by the forest canopy (Hedin et al. 2003).

One possible explanation for the pattern in fig. 7.7a is that both inputs and outputs of Cl are accurately measured and calculated, but Cl accumulates within young sites. This possibility is most unlikely; these systems would have to accumulate more Cl than N, and for thousands of years longer. An alternative possibility is that calculated inputs are accurate in the 0.3 ky site (where atmospheric inputs were in fact measured), but calculated outputs there are inaccurate because the lysimeters sample an unrepresentative fraction of the water moving through the soil. Although soil texture does change progressively across the age gradient (Lohse 2002), I don't think that unrepresentative sampling can explain low Cl losses in young sites. Where streams exist, Cl concentrations in streams follow the same pattern as in lysimeters, and at about the same concentrations (fig. 7.8), suggesting that lysimeters and streams sample similar fractions of hydrologic losses from these sites.

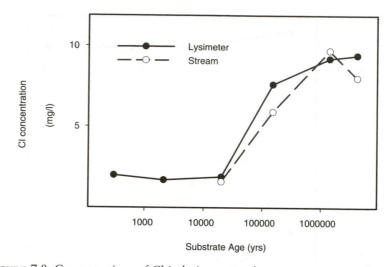

FIGURE 7.8. Concentrations of Cl in lysimeters and streamwater across the substrate age gradient, modified from Hedin et al. (2003); no streams are present near the two youngest sites on the gradient. Cl concentrations are lower in the younger sites, and very similar in streams and lysimeters.

I believe that the most reasonable explanation for the pattern in fig. 7.7a is that atmospheric inputs of Cl are lower in the young than the old sites, despite some inputs of volcanic Cl. A plausible mechanism is that the younger sites are on larger mountains and generally farther from the ocean than the old sites. On smaller mountains, the tradewinds are lifted up, condense moisture, and move on; however, on the larger mountains much of the precipitation results from interactions of air masses moving up- and down-slope (fig. 2.8). Consequently, sites on large mountains are at a greater effective distance from the ocean, and concentrations of sea salt should be lower in precipitation and cloudwater there (Hedin et al. 2003). Consistent with that explanation, precipitation collected near the 4100 ky site yielded higher concentrations of Cl than concurrent precipitation near the 0.3 ky site (Carrillo et al. 2002).

Sulfate too is a mobile anion derived in part from marine aerosol—although in contrast to Cl, SO_4 is affected more by volcanic activity, influenced more by anion exchange and adsorption in soils, and in greater demand by organisms. As with Cl, inputs and outputs of SO_4–S agree closely in the older sites (fig. 7.7b). However, outputs from the 0.3 ky site are greater than measured inputs. Calculated inputs to the 0.3 ky site include a substantial contribution of volcanic S (fig. 6.8b); outputs probably reflect even greater volcanic influence, due to the unmeasured absorption

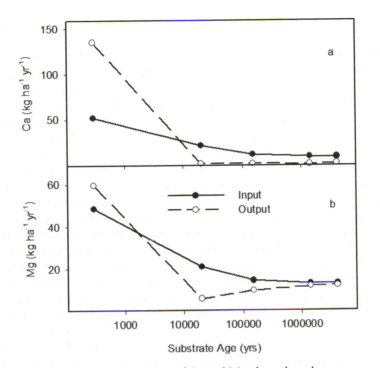

FIGURE 7.9. Total inputs and outputs of Ca and Mg along the substrate age gradient. (a) Ca inputs (from fig. 6.9a) decrease from young to old sites due to declining basalt weathering; outputs (Hedin et al. 2003) decline more dramatically, to very low levels in the old sites. (b) Mg follows a pattern similar to Ca, although with inputs and outputs closer to being balanced throughout the gradient.

of SO_2 gas by the forest (Hedin et al. 2003). High concentrations of SO_2 over the site are episodic, being associated with winds from active volcanic vents on short time scales (Carrillo et al. 2002), and with the occurrence of active eruptions on longer ones.

CALCIUM AND MAGNESIUM

Basalt weathering is a large source of divalent cations in young sites. When that source is depleted, cations are supplied at a lower rate, mostly by atmospheric deposition (fig. 6.9). Consistent with this pattern in inputs, outputs of Ca and Mg are substantially greater in the 0.3 ky site than older ones (fig. 7.9). However, input/output budgets show that Ca in particular is not close to being in balance in any site (fig. 7.9a); outputs far exceed inputs in the 0.3 ky site, while inputs exceed outputs in all of the older sites. Mg is not so dramatically out of balance (fig. 7.9b), except in the 20 ky site where inputs of all elements via marine aerosol might be overestimated.

How can leaching losses of Ca from the 0.3 ky site be so much larger than total Ca inputs? The large losses of SO_4 here (fig. 7.7b) suggest a possible mechanism. Our weathering calculation is based on integrated rates of disappearance of Ca from soils, but leaching was measured for a 2+ year period. Although our calculation of Ca inputs via weathering integrates both active and quiescent phases of Kīlauea Volcano, all of our measurements of outputs occurred during an active eruptive phase when inputs and losses of SO_4—and so of Ca—have been unusually large.

The excess of Ca inputs over outputs in the older sites is more difficult to explain. Atmospheric deposition of both precipitation and cloudwater includes a substantial contribution of non-sea-salt (NSS) Ca, while NSS Ca in lysimeters is negative—meaning that concentrations are below those expected based on Cl concentrations and Ca/Cl ratios in the ocean. (We use Cl as the baseline in lysimeters because weathering contributes to Na losses). I do not know why atmospheric deposition is substantially enriched in Ca (relative to marine aerosol), while lysimeters are substantially depleted. This pattern depends on solution chemistry, not on accurate measurements of water flux. Of course, either input or output chemistry (or both) could be in error—but I doubt that; there have been careful internal and external checks on both (Carrillo et al. 2002, Hedin et al. 2003), and in any case Ca is not readily contaminated or consumed relative to other elements. If both sets of measurements are accurate, their logical consequence is a sink for Ca within the older sites—but there is no evidence for such a sink, and no reasonable way that Ca could accumulate at this rate for hundreds of thousands of years.

In contrast to lysimeters, streamwater Ca concentrations are greater, and include some NSS Ca in all of the older sites—presumably reflecting weathering along the flowpath from lysimeters to streams (Hedin et al. 2003).

SODIUM AND POTASSIUM

Even in young sites, the largest source of Na should be atmospheric deposition (fig. 6.10a). Although plants accumulate K, biotic accumulation of Na is small, and inputs and outputs of Na should balance reasonably well. In fact, calculated inputs of Na far exceed outputs in the 0.3 ky and 20 ky sites; and inputs and outputs balance reasonably well in the older sites (fig. 7.10a)—the same pattern as for Cl^- (fig. 7.7a). This similarity reinforces the suggestion that atmospheric deposition of marine aerosol is overestimated in the young sites.

In contrast, calculated K inputs exceed outputs in young sites as well as older ones (fig. 7.10b). Accumulation in growing biomass in the 0.3 ky

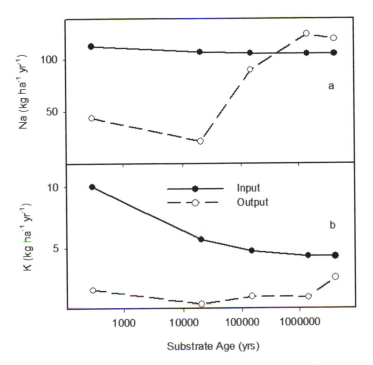

FIGURE 7.10. Total inputs and outputs of Na and K along the substrate age gradient. (a) Na inputs (from fig. 6.10a) and outputs (Hedin et al. 2003) follow a pattern much like that of Cl (fig. 7.7a); both are derived primarily from atmospheric deposition of marine aerosol. (b) Inputs of K are substantially greater than calculated outputs across the entire gradient.

site could account for some of this imbalance—although element inventories in other *Metrosideros* forests (Mueller-Dombois et al. 1984) suggest that no more than 300 kg ha^{-1} of K could be stored over the ~300 years that this site has been accumulating biomass, versus a calculated excess of 8 kg ha^{-1} yr^{-1} of inputs over outputs. In the older sites, K outputs are extremely low—and as with Ca, K concentrations in lysimeters are below the K/Cl ratio of seawater.

SILICON

Unlike Cl, SO$_4$–S, and the major cations, Si is nearly absent from marine aerosol; its major source is basalt weathering, with a contribution from continental dust that increases in relative importance in the oldest sites (fig. 6.11). Moreover, little Si accumulates in biomass, and SiO$_4$–Si is mobile through most tropical soils. To a greater extent than any other element, the budget of Si should reflect a balance between inputs via basalt

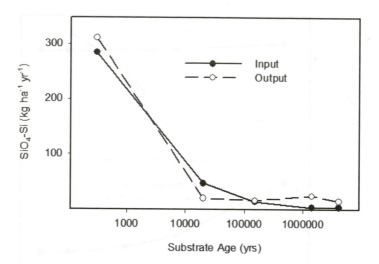

FIGURE 7.11. Total inputs and outputs of SiO_4–Si along the substrate age gradient. Si inputs (fig. 6.11) are dominated by basalt weathering in young sites, and decline substantially in older sites; outputs (Hedin et al. 2003) match inputs very closely across the gradient.

weathering and outputs via leaching. In fact, inputs and outputs of SiO_4–Si balance remarkably well (fig. 7.11), with large inputs and outputs in the 0.3 ky site and much lower fluxes in older sites.

NITROGEN AND PHOSPHORUS

The supply of N demonstrably limits plant growth and NPP in the 0.3 ky site (fig. 5.1), as it does in many of Earth's ecosystems. Consequently, N should be retained strongly in that site, with inputs greater than outputs, and be closer to in balance in the older sites. In fact, calculated inputs of N are much greater than outputs in the youngest site, and surprisingly close to balance thereafter (fig. 7.12a). Streamwater concentrations of NO_3–N also are relatively high (0.08–0.12 mg NO_3–N /liter) near three of the four older sites—excepting the 150 ky site—but they are less than those in lysimeters (Hedin et al. 2003). This difference could reflect hydrological flow paths in soils (Lohse 2002), plant uptake of NO_3–N below the depth of the lysimeters, and/or denitrification or adsorption along the flowpath between lysimeters and streams (Hedin et al. 1998). Depending on the mechanism involved, the difference between lysimeters and streams could exaggerate the agreement between inputs and outputs in the older sites (fig. 7.12a)—but by any measure it is clear that N losses are greater, and closer to inputs, in the older, N-rich sites.

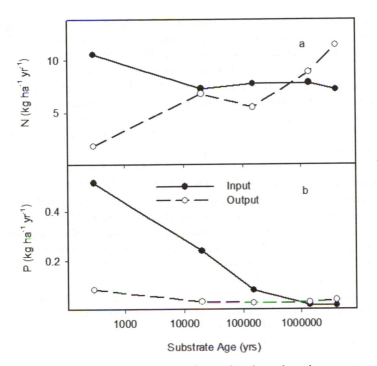

FIGURE 7.12. Total inputs and outputs of N and P along the substrate age gradient. (a) N inputs (fig. 6.12a) vary little, but outputs (from fig. 7.3a) are low in the young site and increase to more or less balance inputs in the older sites, where N availability is greater. (b) P inputs (fig. 6.12b) decline across the sequence, reflecting decreasing inputs via basalt weathering; outputs (fig. 7.3b) are less than inputs in all but the oldest sites.

Calculated inputs of N are ~9 kg ha^{-1}yr^{-1} greater than calculated outputs in the 0.3 ky site, suggesting that this much N accumulates in the developing soils and biomass of that site. If this accumulation was constant over the ~300 yr history of the site, then about 2700 ky of N should have accumulated in plants and soils. In fact, the soils contain 9800 kg of N per ha (Crews et al. 1995), and plants probably contain another ~400 kg of N (Mueller-Dombois et al. 1984). Perhaps the soil in this site has been accumulating N for 500 rather than 300 years (Clague et al. 1999; chapter 3), reducing the discrepancy a little; perhaps Carrillo et al. (2002) higher estimates of atmospheric deposition are more accurate than the values I used. In any case, balancing the N budgets of developing ecosystems in notoriously difficult; it is not uncommon to find more N accumulating than can be accounted for by known inputs (Allison 1955, Bormann et al. 1993, Jenkinson et al. 1994, James 2000—but see Binkley et al. 2000).

The supply of P limits plant growth in the oldest site on the age gradient (fig. 5.1), and in many other ecosystems on old, deeply leached soils (Sanchez 1976, Richter and Markewicz 2001). Inputs of P are greater than outputs in the younger Hawaiian sites (fig. 7.12b)—a surprising observation, in that weathering is measured by the disappearance of P (relative to Nb) from the soil profile, and only plant biomass represents a sink for P that is not accounted for within the weathering calculation. Other *Metrosideros* forests contain 35–40 kg/ha of P in biomass (Mueller-Dombois et al. 1984), while inputs to the 0.3 ky site exceed outputs by ~0.4 kg ha^{-1} yr^{-1}; thus it would require only ~100 years for all of the P in plant biomass to accumulate. However, the calculation of P input via weathering is sensitive to assumptions concerning initial parent material chemistry, particularly in young sites. In the 20 ky and 150 ky sites, I suspect that much of the P input via weathering occurs below the depth of lysimeters, contributing to the excess of inputs over outputs there (fig. 7.12b)—and also contributing to greater concentrations of P in streams relative to lysimeters (Hedin et al. 2003).

In contrast to N, there is a small net loss of P in the 4100 ky site (fig. 7.12b), where P supply demonstrably limits forest growth. As discussed in chapter 8, I believe that P limitation persists in this site because losses of P offset the very low P inputs there.

Using These Element Budgets

These calculations of element input-output budgets differ from those developed in watershed studies in that we integrate some inputs over much longer time scales, and also calculate element inputs via weathering more independently. However, the hydrologic balances of these Hawaiian sites are not nearly so well constrained as those in well-designed watershed studies (Likens et al. 1977, Swank and Crossley 1988)—and this lack of hydrologic constraint is frustrating at times. For a couple of the elements (notably K), the budgets I calculate are far from being in balance—and I am far from understanding why. Nevertheless, even at their worst these budgets make clear what I still don't understand about Hawaiian ecosystems. The most surprising and interesting of these unexplained phenomena is the apparent accumulation of Ca and K in older sites (figs. 7.9a, 7.10b). I believe that the observations underlying these budgets are solid—but I don't know how or where or why Ca and K accumulate, or for how long. Although I have explained the biological dynamics of Hawaiian ecosystems largely in terms of N and P availability, these Ca and K budgets suggest that there remains interesting and potentially important biogeochemistry here that needs to be understood.

Chapter Eight

ISSUES AND OPPORTUNITIES

IN THIS FINAL CHAPTER, I return to the questions raised in chapter 1. How do biological and geochemical processes that operate on very different time scales interact to cause, sustain, or offset nutrient limitation? Can different pathways of element losses from terrestrial ecosystems differentially regulate nutrient cycling and limitation? How do the cycles of different elements interact? How can genetically different populations of organisms, different species, and different levels of biological diversity interact with physiological and geochemical processes to control nutrient cycling and limitation?

INTERACTIONS OF TIME SCALES

Initially, I considered processes controlling nutrient cycling and limitation at four time scales: (1) supply versus demand for nutrients, over time scales of minutes to years; (2) feedbacks between nutrient availability, plant growth, and nutrient supply, over months to decades; (3) sources versus sinks for nutrients, over years to centuries; and (4) inputs versus outputs of nutrients, over centuries to millions of years. In this section, I illustrate the importance of each of these sets of processes across the Hawaiian substrate age gradient, and explore ways in which they reinforce and/or offset each other in positive and negative feedbacks. During these analyses, other sets of processes emerged as being important—notably the replacement of genotypes and species, over decades to centuries—and they will be discussed later in the chapter.

An Exploratory Model

Interactions and feedbacks among processes on time scales ranging from minutes to millions of years are too complex to handle with verbal (conceptual) models, so I developed a simple simulation model that I call "Explore." This model is not intended as a simulation of Hawaiian ecosystems, or any others; rather, it is a way to make my assumptions about nutrient cycling and limitation on multiple time scales explicit, and to examine their consequences. The model represents two pools of C (plant and soil) and three of a nutrient (plant, soil, and available). I had nitrogen in mind while writing the model, so I will call the nutrient "N." However, with a few changes in nutrient use efficiency and critical C:nutrient ratios, the model could be made to fit other elements.

In the model, decomposition is calculated as a rate constant (initially fixed) times the soil C pool. Mineralization of soil N is calculated by assuming that the (implicit) decomposer community has a critical C:N ratio (here 20), below which decomposition releases soil N into the biologically available pool. N uptake from the available pool is calculated by assuming that plants acquire all of the available N, up to a maximum level set by the availability of other (implicit) resources. N that remains in the available pool after plants take up what they can is lost via leaching. Plant production (in units of C) is calculated as N use efficiency (initially fixed at 50 gC/gN) times N uptake. I assume the residence time of C and N in plants to be four years, and that C and N lost to plants via litterfall goes directly into soil C and N. Although it is extremely simple, this model can be modified to incorporate processes at all of the time scales of interest, and to explore their interactions. I will introduce these modifications as I discuss each set of processes. The model and modifications are available as a MATLAB program at www.stanford.edu/vitousek/princetonbook.html.

Supply versus Demand

Proximately, nutrient limitation results from an imbalance between the supply of a nutrient and demand for that nutrient by organisms. Where demand exceeds supply, the available pool is drawn down, and the growth of organisms that require the nutrient can be constrained. A sustained supply-demand imbalance is sufficient to account for nutrient limitation on short time scales.

In the field, we documented supply-demand imbalances on the extremes of the Hawaiian age gradient, for N in the youngest site and P in the oldest one. In the 0.3 ky site, the pool of available N is small (fig. 4.5), and ^{15}N labeling studies show that it turns over rapidly (Hall and Matson 2003). The supply of N by mineralization is low in comparison to other sites (fig. 4.5), and fertilization with N greatly increases the growth of trees and overall ANPP (figs. 5.1, 5.3). Similarly, the pool size of available P is small in the 4100 ky site (fig. 4.4b), ^{32}P labeling studies show that it turns over in minutes (Olander 2002), and plant growth there is stimulated by added P (figs. 5.1, 5.3).

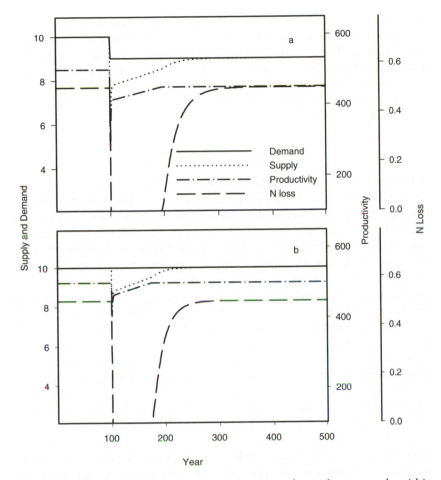

FIGURE 8.1. Consequences of an excess of nutrient demand over supply within the Explore model. (a) Decomposition was reduced by 20% (reducing nutrient supply) and maximum N uptake by 10% (reducing demand) in year 100 of a simulation. (b) 50 units of C/yr were added to the soil beginning in year 100, thereby reducing nutrient supply by increasing immobilization. In both cases, N demand remained less than N supply for ~100 yrs; during this period, productivity was reduced a little and N loss reduced a lot.

To simulate a supply-demand imbalance in Explore, I ran the model to equilibrium (if only that could be done with ecosystems . . .), and then introduced an excess of nutrient demand over supply in two ways. First, I reduced both decomposition and productivity (maximum potential N uptake), the former a little more than the latter. That change reduces nutrient supply slightly more than it does demand. Alternatively, I add 50 units of extra C to the soil pool each year, thereby raising the C:N ratio and (in effect) causing microbes to reduce net N mineralization by immobilizing N. Either way, simulated plant growth becomes N limited, and the available pool of N drops to zero after plants have taken up what they can (fig. 8.1). Because I assume that losses of N originate in the available pool, the disappearance of available N shuts off N outputs from the system as a whole.

Plant-Soil-Microbial Feedbacks

While an excess of nutrient demand over supply proximately causes nutrient limitation, it doesn't explain limitation. For biophilic nutrients—those primarily associated with organisms and organic matter—plant and microbial growth and turnover drive nutrient demand, while decomposition and nutrient mineralization drive supply. Microbial decomposers thus play a dual role as both suppliers and demanders of nutrients. However, the stoichiometry of plants is extremely C-rich relative to that of microbes—as expressed in the Explore model with a plant C:N ratio of 50 versus an implicit microbial C:N ratio of 10. (The critical element ratio of 20 for N mineralization is based on a microbial C:N ratio of 10 and a microbial growth efficiency of 50%.) Plant litter therefore represents an N-poor substrate for microbes, causing them to immobilize N during decomposition. Moreover, plants in low-nutrient sites generally have greater nutrient-use efficiency; they retain nutrients longer within their tissues (Berendse and Aerts 1987) and produce more recalcitrant and/or toxic compounds that defend those tissues against herbivores (Rhoades 1979, Reich et al. 1997). Their litter decomposes and regenerates nutrients more slowly than does litter produced in more fertile sites (chapter 4), leading to a positive feedback from low nutrient availability in soil to efficient nutrient use in plants to low rates of nutrient mineralization (fig. 4.17) (Vitousek 1982, Pastor and Post 1986, Wedin and Tilman 1990, Hobbie 1992). Where it occurs, this feedback clearly reduces nutrient supply—and it also reduces the demand for nutrients in infertile sites, because efficient plants require less of a nutrient to make the same amount of growth. Whether this plant-soil-microbial feedback drives nutrient limitation per se depends on the relative magnitude of the reduction in nutrient demand by plants versus supply by microbes.

I incorporated a plant-soil-microbial feedback into the Explore model by varying plant nutrient use efficiency (NUE) as a function of actual N uptake relative to maximum N uptake, and by varying the rate of organic matter decomposition as a linear function of NUE. As uptake declines from the maximum potential uptake to half the maximum level, I increased NUE from 50 to 90—causing a slow decline in productivity. If

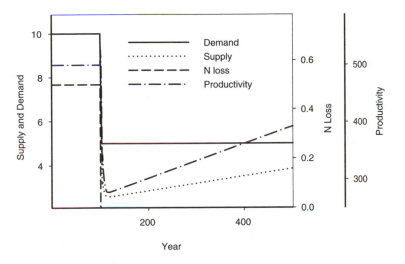

FIGURE 8.2. The consequences of incorporating a dynamic plant-soil feedback into the Explore model. When N supply is less than demand (beginning in year 100, with the addition of 50 units/yr of C to the soil), plant N use efficiency is increased (thereby increasing litter C:N ratio) and decomposition rate is decreased. The greater nutrient use efficiency of plants decreases demand for N, but the slower decomposition and higher C:N ratio decrease supply to a greater extent. The feedback increases both the intensity and duration of an imbalance between nutrient supply and demand.

uptake declines to < 50% of the maximum, NUE remains 90—and productivity declines sharply. These changes reduce both nutrient supply and demand, and they allow a plant-soil-microbial feedback to develop where N supply is reduced relative to demand.

One lovely feature of models is that unlike ecosystems, they can be evaluated in the presence and the absence of a process. Fig. 8.1 illustrates the consequences of decreasing nutrient supply by adding C to soils when the plant-soil-microbial feedback is turned off. Adding C with the feedback turned on, so that nutrient use efficiency and decomposition are allowed to vary, decreases supply more than it does demand, and increases both the intensity and the duration of N limitation to productivity (despite the greater NUE of plants) (fig. 8.2). This result is similar to that of Tateno and Chapin's (1997) model, when they evaluated an open system with a variable rate of decomposition.

Note that plants and decomposers in this version of the model respond immediately to changes in nutrient availability. How does this aspect of the Explore model compare with *Metrosideros*-dominated forests in Hawai'i?

Changes in both NUE and decomposition along the age gradient are substantial, and comparable to those in the model (chapter 4). However, responses to fertilization (chapter 5) demonstrate that while both P-use
efficiency and decomposition change significantly and relatively rapidly
in response to P fertilization in the P-limited 4100 ky site, neither N use
efficiency nor decomposition respond substantially to N fertilization in
the N-limited 0.3 ky site (fig. 5.8). As discussed in chapter 5, the *Metrosideros* population that dominates the N-limited site differs genetically
from populations in older sites (Treseder and Vitousek 2001b); this population might have to be replaced before any plant-soil-microbial feedback
initiated by fertilization can be expressed. The timing and implications of
genotype and species replacements will be discussed later in this chapter.

Sources and Sinks

Sinks are pools of nutrients that accumulate over time; in effect, nutrients
in the sink are removed from circulation in the ecosystem as a whole for
as long as that pool remains intact. One well-characterized sink is the accumulation of nutrients in the growing biomass (including litter and soil
organic matter) of successional forests (Gorham 1961, Vitousek and Reiners 1975, Bormann and Likens 1979). By removing nutrients from circulation, this sink can reduce nutrient availability at a critical stage of stand
development (Miller 1981, Pastor and Post 1986). On the Hawaiian age
gradient, the accumulation of N in the soil and trees of the N-limited
youngest site represents a substantial sink, one that is greater on an annual
basis than measured N inputs (fig. 6.12a). Beyond early primary succession,
however, nutrient sinks in Hawaiian forests are relatively poorly defined.
Disturbance caused by stand-level dieback of *Metrosideros* (Mueller-
Dombois 1986, Akashi and Mueller-Dombois 1995) and by relatively
infrequent but intense hurricanes (Herbert et al. 1999) has been characterized, but observations of subsequent forest regrowth and nutrient accumulation are sparse. Moreover, the absence of reliable annual tree
rings makes it difficult to reconstruct the past dynamics of forest growth
and turnover with the level of precision that can be attained in many
higher-latitude forests (Foster 1988, Fastie 1995). Much as it hurts me to
say it, the Hawaiian Islands are not a good model system for understanding nutrient source-sink dynamics associated with disturbance and
forest regrowth.

I modified the Explore model to simulate a nutrient sink in growing
biomass by adding pools representing wood C and N; these pools turn
over much more slowly than the rest of plant tissue, every 50 years rather
than every four years. Although wood has a much wider C:N ratio than
other plant tissues, I can already explore the implications of varying C:N
ratios (fig. 8.2)—so to avoid confounding sinks with feedbacks, I set the

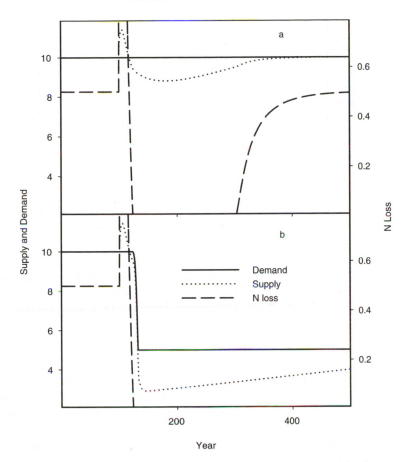

FIGURE 8.3. The consequences of adding a nutrient sink in growing biomass to the Explore model. Beginning in year 100, 20% of C and N uptake was allocated to an accumulating slow-turnover pool intended to simulate wood. (a) N supply and demand and N losses with the plant-soil-microbial feedback off. (b) N supply and demand and N losses with the feedback on. The consequences of adding a nutrient sink in growing biomass are similar to those of creating an excess of demand over supply.

C:N ratio of wood equal to that of other plant tissues. When the wood portion of the model is turned on, 20% of C and N uptake are allocated to wood. I ran the model to equilibrium without wood, then initiated the growth of wood at year 100—as might occur if woody plants invaded a perennial grassland. With the plant-soil-microbial feedback off, wood growth creates a sink for N that accumulates for decades; after a brief transient period in which the lower return of C and N in litterfall causes a pulse of nutrient availability and loss, this sink reduces nutrient supply

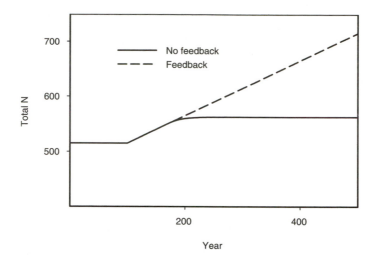

FIGURE 8.4. The simulated N sink in soil caused by an excess of nutrient demand over supply (induced by adding C to soil) is larger and sustained much longer when the plant-soil-microbial feedback is on (within the model) than when it is off.

relative to demand and greatly reduces N outputs (fig. 8.3a). With the feedback on, these effects are more intense and longer-sustained (fig. 8.3b).

In addition, simply adding C to soil creates a demand-supply imbalance that persists until enough additional N accumulates to restore the original rate of N mineralization. Even when the plant-soil-microbial feedback is off, this accumulation represents an N sink in soil that continues for decades. With the feedback on, the sink for N in soil can continue to accumulate for centuries (fig. 8.4).

Any process that destroys or decreases accumulated pools of nutrients within ecosystems turns what had been nutrient sinks into sources of nutrients. Disturbances (fires, clearcuts, hurricanes, etc) can increase nutrient supply by returning accumulated nutrients from plants to soils and by increasing soil temperature, thereby increasing decomposition/mineralization; they also temporarily decrease plant demand for nutrients. Consequently, disturbance often causes a substantial pulse in the availability and loss of nutrients (Likens et al. 1970, Bormann and Likens 1979, Vitousek and Melillo 1979, Matson and Boone 1984, Vitousek and Matson 1985), with implications that will be discussed later in this chapter.

Inputs and Outputs

Growing vegetation and developing soil can be substantial sinks for nutrients for decades to centuries—and while those sinks accumulate, nutrient supply can limit biological processes, and inputs of nutrients to ecosystems should be greater than outputs. However, neither plant biomass nor soil organic matter can accumulate nutrients without limit; the sinks that they represent eventually must either fill up or be reset by disturbance. If those sinks fill, then nutrient outputs must increase to approach inputs, or inputs must decrease, or both. This balance between inputs and outputs represents the final time scale in my analysis.

A number of observations on the Hawaiian age gradient illustrate the importance of input-output balances in controlling nutrient availability. For example, the high rates of Ca and Mg input via basalt weathering in the 0.3 ky site lead to relatively high cation availability there (figs. 4.4a, 6.9). Also in that young site, inputs of N far exceed measured outputs (fig. 7.12a), consistent with an accumulating sink for N in soil (fig. 4.3)—but older sites have little or no net N sink in soil, greater N availability, and substantial N losses (figs. 4.3, 4.5, 7.12a). In the oldest site, losses of P via extremely dilute concentrations of inorganic and organic P in percolating water offset or even exceed the very low inputs of P via basalt weathering and continental dust (fig. 7.12b)—thereby causing and/or sustaining P limitation, as discussed below.

In the Explore model, I treat N outputs as originating in the biologically available pool of N in soil, and occurring only after plants have acquired all the N that they can—corresponding reasonably well with N losses via NO_3–N leaching. In contrast, N inputs are independent of ecosystem properties, and constant—corresponding to inputs of fixed N dissolved in precipitation. Wherever simulated nutrient demand is greater than supply, all of the available N is taken up by plants, so N losses are zero and inputs of N accumulate within the modeled system.

The modeled responses to supply-demand imbalances, plant-soil feedbacks, and source-sink dynamics illustrated in figs. 8.1 to 8.4 depend upon N inputs, as can be demonstrated by setting inputs to zero within the model. I started Explore at equilibrium, then imposed an excess of nutrient demand over supply by adding C beginning at year 100. Without inputs, the modeled system quickly jumps into a low-supply, low-demand, N-limited mode (fig. 8.5b,d)—and stays there indefinitely. However, in an open system, inputs of N eventually accumulate to the point where even a slow rate of turnover of a large pool of soil N can supply all the N that plants require—whereupon the plant-soil-microbial feedback reverses direction, driving the system towards low efficiency, rapid decomposition, and more rapid N cycling (fig. 8.5a,c). Indeed, after 800 years (for the conditions simulated in fig. 8.5), the feedback no longer affects N cycling or limitation at all. As long as inputs and outputs are controlled as modeled here, the accumulation of nutrient inputs represents a long-term negative feedback to nutrient limitation.

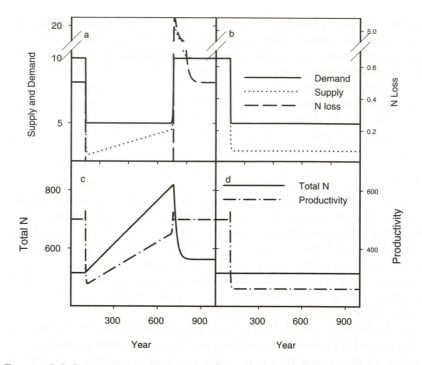

FIGURE 8.5. Long-term consequences of nutrient input-output balances within the Explore model. Simulations were run for 1000 years with the plant-soil-microbial feedback active; an excess of nutrient demand over supply was created by adding 50 units of C beginning in year 100. (a, c) As in fig. 8.2, a supply-demand imbalance initiates a plant-soil-microbial feedback and reduces N supply and demand, N losses, and productivity; the resulting excess of nutrient inputs over outputs causes N to accumulate within the system. After ~700 years, enough N has accumulated so that its slower turnover nevertheless supplies sufficient N for plant growth; at this point the feedback reverses direction, towards an excess N supply and high losses, until the system equilibrates close to where it started. (b, d) Without inputs of N, nutrient cycling in the simulated system remains stuck in a low-supply, low-demand, low-productivity N-limited state.

Interactions across Scales

One striking result of the model is that nutrient supply and demand, plant-soil-microbial feedback, and nutrient sources and sinks—the three fastest sets of processes—all reinforce each other in positive feedbacks that reduce nutrient availability in nutrient-poor sites and enhance it in rich sites. Inducing an excess of nutrient demand over supply causes nutrient limitation to plant production (fig. 8.1) and a sink for N in soil (fig. 8.4);

the responses of plants and decomposers to nutrient limitation sets a plant-soil-microbial feedback in motion, increasing and extending the excess of nutrient demand over supply (fig. 8.2) and creating a larger sink for N in soils (fig. 8.4). Similarly, introducing an N sink in growing biomass into the model causes an excess of nutrient demand over supply, and therefore causes nutrient limitation to plant production; nutrient limitation in turn initiates the plant-soil feedback and creates an N sink in soil (fig. 8.3). As discussed below, understanding the responsiveness of the plant-soil-microbial feedback to nutrient limitation—particularly the extent to which its full operation depends on the replacement of the genotypes or species occupying a site, versus phenotypic responses within those genotypes or species—calls for considerably more work. Nevertheless, I believe that positive feedbacks among processes occurring at very different time scales are important in the real world of plants, microbes, and soils as well as in the electronic world of the model.

As long as I assume that nutrient inputs are constant, and that nutrient losses occur from the biologically available pool, then any change in the balance between supply and demand will also affect the balance between nutrient inputs and outputs. Where nutrients are in relatively short supply, all three of the faster sets of processes interact to reduce supply relative to demand—and so to reduce N losses, causing inputs to be greater than outputs. In this sense, the more rapid processes control the slowest set—but the slower processes have relatively little immediate effect on the faster ones, because annual nutrient inputs generally are small relative to the quantities cycled and/or held within ecosystems. However, if nutrient inputs are greater than outputs for long enough, the consequent accumulation of nutrients eventually suffices to fill any sink—reversing the direction of the plant-soil-microbial feedback, and eventually aligning nutrient demand with nutrient supply (fig. 8.5). Nutrient input-output balances thereby provide a long-term negative feedback to nutrient limitation, one that can offset all of the faster processes (supply and demand, plant-soil feedback, sources and sinks).

The time scales involved in nutrient input/output balances can be very long. For example, I estimate total annual inputs of N and P into the 4100 ky site on the Hawaiian age gradient to be 7.2 and 0.016 kg ha^{-1} yr^{-1} respectively (fig. 6.12), versus total pools of N and P in soils of 11,300 and 4670 kg/ha (Crews et al. 1995). Consequently, the mean residence times for total N and P are 1570 and > 280,000 yrs respectively. This residence time for P is more than twice as long as a complete glacial-interglacial cycle. Glaciation represents a boundary in the past of many temperate and boreal ecosystems—a starting point from which they developed. However, the dynamics of Hawaiian and other tropical and subtropical ecosystems can be influenced by biogeochemical processes that occurred under full-glacial climates and plant communities very different from those of the present (Hotchkiss 1998, Hotchkiss et al. 2000). Nevertheless, even the residence time of P is short relative to the age of the 4100 ky site—and most biologically-essential elements turn over much more rapidly than does P. Why then does nutrient supply limit biological processes over so much of Earth? Why doesn't the negative feedback provided by the accumulation of nutrient inputs to open systems eventually align nutrient supply with demand, source with sink, and input with output, all at a level where nutrient supply does not limit biological processes?

The Regulation of Nutrient Inputs and Outputs

Although the Explore model assumes that nutrient inputs are constant, in fact most pathways of element input are affected by features of the ecosystems that they enter. For example, canopy geometry affects cloudwater interception (Clark et al. 1998, Weathers et al. 2000), and biological activity increases soil acidity and therefore enhances rates of weathering (Cochrane and Berner 1997). However, while element inputs by most pathways are influenced by ecosystems, they are not regulated by ecosystems—in the sense that demand for a particular element does not enhance inputs of that element. Moreover, where inputs are responsive to demand—as certainly occurs with biological N fixation (Hartwig 1998), and as could occur where P- or Ca-deficient microbes enhance the weathering of apatite (Newman and Banfield 2002, Blum et al. 2002)—such inputs would accentuate the negative feedback between nutrient limitation and the input-output balances of ecosystems, making sustained nutrient limitation even more difficult to explain. It is just this consequence of biological N fixation that makes the widespread occurrence of N limitation to primary production such an intriguing puzzle (Vitousek and Howarth 1991, Vitousek and Field 1999).

In contrast to inputs, the Explore model treats nutrient outputs as representing the residual pool of available nutrients that remains after organisms have taken up what they can. Outputs by this pathway can regulate the quantity of nutrients in ecosystems, in the sense that an excess of nutrient demand over supply reduces outputs and leads to nutrient accumulation, while an excess of supply over demand causes nutrient losses in excess of inputs, and so decreases the quantity of a nutrient within an ecosystem. This pathway of element loss is an important one in many ecosystems—but as I discuss in chapter 7, there are additional pathways of element loss that are controlled differently, and that could have different effects on element input-output budgets. Although the loss of dissolved organic N (DON) represents the best-characterized of these pathways

(Hedin et al. 1995, Perakis and Hedin 2001, 2002; chapter 7), several additional pathways of nutrient loss share with DON the feature that losses can persist even when organisms are strongly nutrient-limited. In chapter 7, I referred to these losses as representing elements that organisms cannot get; here I describe them as demand-independent losses, because they continue even when the demand for an element exceeds the supply of that element.

Demand-Independent Pathways of Element Loss

Pathways of element loss that might be at least partially demand-independent include losses that occur during and following disturbance, losses resulting from the asynchrony of nutrient demand and supply, and losses via erosion and the long-term adsorption and/or occlusion of nutrients in soils—in addition to the pathways discussed in chapter 7, the leaching of dissolved organic forms of nutrients, and of minimum concentrations of available nutrients. I will describe each of these pathways briefly here, and then discuss how they could interact to sustain nutrient limitation to primary production and other biological processes.

DISTURBANCE

Some forms of destructive disturbance (such as biomass harvests and fires) remove nutrients from ecosystems, rapidly resetting biotic pools to a point where nutrient accumulation in regrowing biomass represents a sink that can drive continued nutrient limitation. In addition, these and other disturbances (e.g., windstorms, insect outbreaks) temporarily reduce the demand for nutrients, causing an excess of nutrient supply over demand and so driving post-disturbance losses. These post-disturbance losses represent excess available nutrients when they occur—but where they are large enough and disturbances frequent enough, such losses can contribute to nutrient limitation during most of the interval between disturbances (Vitousek et al. 1998; fig. 8.6). The importance of disturbance in shaping long-term nutrient budgets is widely recognized, and a number of studies demonstrate that nutrient removals in harvest and/or volatilization in fires can lead to nutrient accumulation that continues for more than a century following disturbance (Aber and Driscoll 1997, Goodale and Aber 2001). The importance of post-disturbance losses is less clear, particularly in low-nutrient sites (Vitousek et al. 1982).

ASYNCHRONY

Nutrient supply via mineralization and the demand for nutrients by plants do not respond identically to seasonal or stochastic variations in temperature and especially rainfall (Likens et al. 1977, Davidson et al.

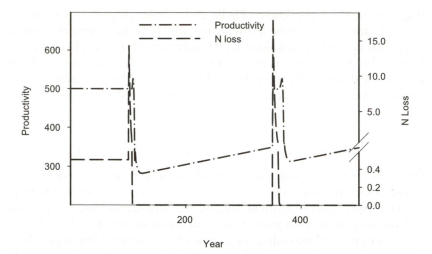

FIGURE 8.6. The effects of destructive disturbance on productivity and N losses in the Explore model. Disturbances in years 100 and 350 remove almost all plant C and N, and cause additional post-disturbance N losses by temporarily reducing inputs of C-rich litter, thereby driving the mineralization and loss of soil N. N losses during and after disturbance in turn cause supply to be less than demand between disturbances, leading to low productivity, low N losses, and N limitation to plant growth in the interval between disturbances.

1993, Burke et al. 1997), and microbial immobilization can further decouple nutrient supply and demand in time. Spatial variation in nutrient availability caused by animal activity or other processes can similarly decouple nutrient supply and demand in space (Vitousek et al. 1998). Consequently, even in a nutrient-limited ecosystem, there can be times and places when nutrient supply is greater than demand. The resulting nutrient losses represent excess available nutrients when and where they occur. However, if such losses are large enough, they can sustain nutrient limitation over most of an ecosystem and/or most of the time—as is illustrated in fig. 8.7, which is based on a model similar to Explore that additionally includes the influence of fluctuating rainfall (Vitousek and Field 2001).

ADSORPTION AND OCCLUSION

Long-term adsorption and/or occlusion (physical protection) of elements in soils represent abiotic sinks that can compete with organisms for available nutrients and so remove those nutrients from the biological portion

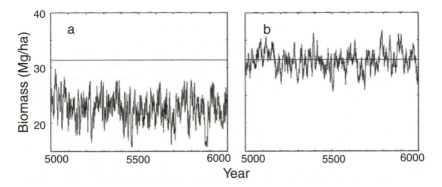

FIGURE 8.7. The consequences of fluctuating soil moisture, based on a simulation similar to Explore (Vitousek and Field 2001). (a) The upper horizontal line represents plant biomass in a system with low but constant rainfall; the jagged lower line is the biomass of a system with the same average rainfall, but substantial year-to-year variation. (b) Biomass in a system with similar rainfall, but with enough N added to reverse N limitation. The lack of change to the horizontal line demonstrates that N limitation does not occur in the constant-rainfall case, while the increase in biomass represented by the jagged line demonstrates that N supply limits biomass accumulation in this fluctuating environment.

of ecosystems. Occlusion contributes substantially to the Walker and Syers (1976) conceptual model for long-term P limitation, as discussed in chapter 4. Across the Hawaiian age gradient, we observed an increased fraction (up to ~50%) of soil P is present in the occluded fraction in the older soils (fig. 4.2, Crews et al. 1995), amounting to a loss of 47 g m^{-2} of P between the 1400 ky and the 4100 ky site (0.00018 kg ha^{-1} yr^{-1}).

EROSION

Where erosion is intense and relatively frequent (as on steep slopes shaped by landslides) it represents a demand-independent pathway of nutrient loss that resets soil development and reinitiates primary succession, thereby sustaining nutrient limitation (Scatena and Lugo 1995, Zarin and Johnson 1995, Walker et al. 1996). Even where it is much less intense, erosion preferentially removes nutrient-rich upper soil horizons (Jobbagy and Jackson 2001)—and for atmospherically-derived nutrients such as N, any such losses must be replaced from outside the system. However, erosion plays a dual role for elements that can be supplied by weathering, both removing elements and rejuvenating the supply of rock-derived nutrients on old substrates (fig. 7.2) (Vitousek et al. in press, Porder et al. submitted).

MINIMUM AVAILABLE CONCENTRATIONS

The Explore model to the contrary, even nutrient-limited organisms can only draw the available nutrient pool in soil down to some non-zero minimum concentration—and leaching of residual nutrient remaining below that minimum represents a demand-independent pathway of nutrient loss. Minimum concentrations vary across elements and ecosystems, reflecting element mobility in soil (Nye and Tinker 1977), allocation to nutrient acquisition (which varies a function of nutrient availability; Rastetter and Shaver 1992, Rastetter et al. 1997, Treseder and Vitousek 2001a), and the properties of individual plant and microbial species. Indeed, the ability of some plant species to draw resources down to lower minimum concentrations than others provides the basis for the R^* concept in plant community ecology (Tilman 1988).

We observed losses of biologically available NH_4–N and NO_3–N from the N-limited 0.3 ky site, and of PO_4–P from the P-limited 4100 ky site (fig. 7.3). Losses of available N from the 0.3 ky site represent only 3.6% of annual N inputs, and N pools continue to accumulate (chapter 7). However, calculated PO_4–P losses from the 4100 ky site are 0.023 kg ha^{-1} yr^{-1}—greater than calculated P inputs to that site, suggesting that these losses could constrain P accumulation and sustain P limitation in the long term (Hedin et al. 2003). Moreover, as discussed below, it is possible that the low diversity and/or idiosyncratic species composition of Hawaiian ecosystems could lead to relatively high minimum nutrient concentrations.

DISSOLVED ORGANIC NUTRIENTS

As discussed in chapter 7, a number of recent studies have demonstrated that high-molecular-weight DON can function as a demand-independent pathway of N losses from terrestrial ecosystems (Hedin et al. 1995, Lajtha et al. 1995, Currie et al. 1996, Perakis and Hedin 2001, 2002). Moreover, both simple and complex ecosystem models demonstrate that if DON losses are large enough (relative to N inputs), they are capable of constraining N accumulation and sustaining N limitation indefinitely (Vitousek et al. 1998, Raich et al. 2000)—as is illustrated for the Explore model in fig. 8.8.

However, DON losses are not large enough to constrain N accumulation and sustain N limitation along the Hawaiian age gradient (fig. 7.3). Losses of DOP are more complex, and probably more important as well. The complexity arises because DOC:DOP and DON:DOP ratios vary with depth and among sites along the Hawaiian age gradient (fig. 7.6), in patterns that suggest that DOP losses are not strictly demand-independent. Nevertheless, the calculated DOP losses of 0.016 kg ha^{-1} yr^{-1} are equivalent to total P inputs in the P-limited 4100 ky site (figs. 7.3b, 7.12b).

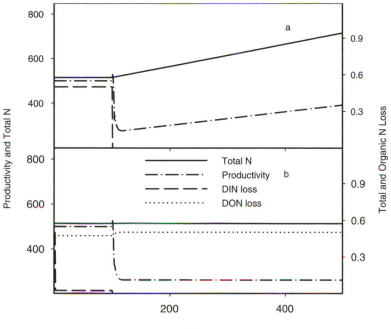

FIGURE 8.8. The effects of N losses via dissolved organic N (DON) within the Explore model. An excess of nutrient demand over supply was introduced in year 100 by adding 50 units of C to the soil. (a) In the absence of DON losses, productivity and losses of inorganic N decline, as in fig. 8.2; soil N accumulates and would eventually reverse the shortage of N. (b) When DON losses are significant (here 0.1% of total soil N per year) and demand-independent, then N inputs cannot accumulate and the system gets stuck in a low-demand, low-supply, N-limited mode. If N inputs to the system were greater, they could overwhelm DON losses and eventually return the system to a more productive state (Vitousek et al. 1998).

Implications of Demand—Independent Nutrient Losses

Clearly, elements can be lost via several pathways that are partially or wholly independent of the overall balance between nutrient supply and demand in terrestrial ecosystems. Losses by these demand-independent pathways are capable of breaking the negative feedback from nutrient limitation to nutrient accumulation, and so capable of sustaining long-term nutrient limitation. The best example of this phenomenon along the Hawaiian age gradient occurs in the P-limited 4100 ky site. I calculate that inputs of P there total 0.016 kg ha^{-1} yr^{-1}, mostly from continental dust and basalt weathering (fig. 6.12b). In contrast, calculated losses are

~2.5-fold greater (fig. 7.3b), with 0.023 kg ha^{-1} yr^{-1} lost as PO$_4$–P, 0.016 kg ha^{-1} yr^{-1} lost as DOP, and a trivial 0.00018 kg ha^{-1} yr^{-1} lost to occlusion. I'm not confident enough in these calculations to conclude that demand-independent losses are truly greater than inputs, and so they deplete P in this site. However, I do think it is reasonable to suggest that P outputs constrain P accumulation there, and so sustain P limitation.

I conclude this section by making three general points concerning demand-independent losses. First, the relationship between nutrient inputs and demand-independent losses is more important than the magnitude of losses per se. For example, demand-independent P outputs probably are at least as large from young sites as from the 4100 ky site on the Hawaiian age gradient, and total P losses are several-fold greater from young sites (fig. 7.3b). However, inputs of P via basalt weathering are much greater than losses in the young sites, and so the availability of P increases during the first tens to hundreds of thousands of years of ecosystem development (figs. 4.4, 6.12b, 7.3b). Basalt weathering is depleted as a nutrient source by the 4100 ky site, and so inputs are much less there—allowing smaller losses of P to place a severe constraint on P accumulation (fig. 7.12b).

Second, very different processes are included in the category of demand-independent losses—and it is the total loss by all such pathways, integrated over space and time, that controls nutrient availability and limitation in the long term. It would be easy to underestimate the importance of demand-independent losses by focusing on one or a few pathways or time scales of loss; what matters is whether the sum of all such losses is large relative to inputs of nutrients.

Finally, with the exception of disturbance and (more recently) DON, our understanding of the ecological importance of demand-independent nutrient losses is not well developed. However, ecosystems in which the budgets of important nutrients are regulated by demand-dependent versus demand-independent losses differ fundamentally (Hedin et al. 1995, Vitousek et al. 1998, Perakis and Hedin 2002)—and where human activities increase nutrient inputs to the point that they overwhelm demand-independent losses, they can profoundly change the nature of an ecosystem (Ågren and Bosatta 1988, Aber et al. 1989; 1998; Fenn et al. 1998). Human-caused changes in disturbance regimes similarly can alter the ways that nutrient cycling is regulated.

STOICHIOMETRY AND FLEXIBILITY

Most analyses of ecosystem-level nutrient cycling include multiple elements but consider them one at a time, separately evaluating the controls of N, P, and cation cycling—as I have done through most of this book. At one level, this is a reasonable approach, because element cycles differ substantially in their sources and controlling mechanisms. However, organisms simultaneously require a full suite of essential elements, and the supply of one element can thereby affect the cycling of many others. One way to incorporate these effects is by evaluating element interactions, the reciprocal influences of element cycles (Bolin and Cook 1983). However, I believe that the concept of stoichiometry provides a more fundamental and useful approach.

The basis of the stoichiometric approach is straightforward (Sterner and Elser 2002). Chemical reactions occur at characteristic ratios of reactants and yield characteristic ratios of products, all of which can be defined in terms of their elemental composition. Moreover, many biochemical reactions are catalyzed by enzymes that themselves have defined elemental compositions, and take place within organisms that have more or less defined compositions. Chemical reactions and organisms both require all of their reactants and catalysts—and in the case of organisms, their structures—if they are to proceed and/or grow. Although the elemental composition of products, reactants, catalysts and structures is an incomplete description of reactions (or organisms), elements are the most conservative component of these reactions. Unlike energy, unlike organic or inorganic compounds, elements are neither created nor consumed—and it is possible to calculate a mass balance for any element in any reaction (other than fission, fusion, or radioactive decay), and for any organism.

Stoichiometric approaches have long been embedded in ecology—for example, they underlie the use of critical C:N ratios in decomposition and nutrient release (Waksman and Tenney 1928). Stoichiometry was applied explicitly on a very broad scale by Redfield (1958), who described relationships among C, N, P, and S in marine algae and bacteria, and consequently among the cycles of C, N, P, S, and O in the ocean. One legacy of his pioneering analysis is that C:N:P ratios in marine phytoplankton are termed the Redfield ratios. Reiners (1986) later described "the stoichiometry of life" as one of the fundamental bases of ecosystem ecology, suggesting that all living organisms could be separated into "protoplasmic life"—which might follow the Redfield ratios—and structural components that can be enriched in particular elements (C for terrestrial plants, Ca and P for terrestrial vertebrates).

More recently, Sterner and Elser (2002) built upon their own and others' research to broaden the scope of the stoichiometric approach substantially, developing and evaluating its implications on levels of organization from organelles to ecosystems. Among many contributions, they analyzed the variability in element ratios within as well as among groups of organisms, demonstrating that the marine phytoplankton discussed by Redfield (1958) have the least variable ratios, and terrestrial plants are the most variable. Much of the variation within groups of organisms is caused by differences in the quantity and biochemistry of structural tissues, as Reiners (1986) suggested, and some is due to storage or "luxury consumption" (uptake in excess of immediate requirements) of elements when they are abundant. However, "protoplasmic life" itself has variable ratios of N:P as well as C:N and C:P, due in part to an association between rapid growth rates and high P concentrations (Elser et al. 1996).

I agree with Reiners (1986) and Sterner and Elser (2002) that together with ecological energetics, biological stoichiometry offers a solid (if incomplete) framework for understanding ecological systems. Further, I believe that linking the biological stoichiometries they discuss with the geochemical stoichiometries exhibited by element inputs to and outputs from terrestrial ecosystems can provide a common language and framework for integrating biogeochemistry as a whole. Finally, I believe that the flexibility of element cycling (Binkley et al. 1992) needs to be considered along with stoichiometry. I define "flexibility" as the ability of a particular element to cycle more rapidly within or between ecosystems than do other elements undergoing similar transformations. When they are in demand, elements with flexible cycles are transformed or transferred more rapidly than would be expected based on the underlying stoichiometry of the organisms and/or processes involved. Flexibility is important because in the long run, it is the element with the least flexible cycle rather than the one that is least abundant (relative to organisms' immediate requirements) that limits biological processes.

In this section (based on Vitousek, 2003), I first analyze the stoichiometry and flexibility of within-system nutrient cycling along the Hawaiian age gradient, focusing on processes that could allow P to cycle through soil organic matter to organisms more rapidly than does N. Next, I evaluate the stoichiometry and flexibility of element inputs to Hawaiian ecosystems. Finally, I discuss constraints to flexibility in the case of biological N fixation—addressing why N limitation to NPP and other ecosystem processes is widespread despite the apparent flexibility of N inputs that fixation provides.

Within-System Element Cycling

At the start of this chapter, I considered nutrient cycling and limitation in terms of the balance between element supply and demand. From the perspective of stoichiometry, organisms never demand one single element; rather, their demand is for energy and a suite of elements in characteristic ratios that vary across groups of organisms (Sterner and Elser 2002). Nutrient cycling within ecosystems can be viewed as a consequence of groups of organisms with very different energy sources (autotroph versus heterotroph, consumer versus decomposer) and characteristic stoichiometries making their livings in the same place—with the tissues and/or waste products of each group being the fundamental resources for some or all of the others.

To date, the stoichiometric approach has been developed more fully in aquatic than terrestrial ecosystems—in large part because the slow turnover and overwhelming abundance of structural and other recalcitrant C compounds in terrestrial ecosystems makes them relatively difficult to evaluate. Nevertheless, as Sterner and Elser (2002) point out, there is a great deal to be gained from applying a stoichiometric perspective to terrestrial ecosystems—especially in that the widest differences in C:element ratios among groups of organisms anywhere occur between terrestrial plants and bacteria, the dominant producers and decomposers of terrestrial ecosystems.

I calculated the ratios of C:N, C:P, and N:P in the leaves, leaf litter, roots, and wood of *Metrosideros* from across the Hawaiian age gradient, using the information summarized in chapter 4. Trees vary substantially across tissues and sites in their C:N and C:P ratios; they are particularly C-rich (N- and P-poor) in the infertile young and old sites on the sequence (figs. 8.9, 8.10), reflecting their greater nutrient use efficiency in those sites. In contrast, N:P ratios vary less across sites and tissues, displaying a relatively consistent stoichiometry (fig. 8.11) despite N limitation to plant productivity in the youngest site versus P limitation in the oldest (Fig 5-1, 5-3).

Leaves of several additional plant species were sampled across the Hawaiian age gradient, again yielding wider variation in C:N and C:P ratios than in N:P ratios among species and sites (fig. 8.12). However, other elements follow quite different patterns. For example, Ca:P ratios vary by a factor of > 6 for *Metrosideros* leaves collected across the Hawaiian age gradient, and by a factor of > 15 among species (fig. 8.13). The cation stoichiometry of ferns in particular differs substantially from that of angiosperms; ferns are low in Mg and especially Ca, and slightly enriched in K. I wonder if ferns' stoichiometry affects their decomposability; fungi in particular require relatively large quantities of Ca (Silver and Miya 2001), and tree fern litter decomposes very slowly despite high concentrations of N and P and low concentrations of polyphenols and particularly lignin in their tissues (Scowcroft 1997, Hättenschwiler et al. 2003).

Element ratios in microbial biomass along the age gradient were measured using chloroform fumigation/extraction (Torn et al. submitted). I am not convinced that this technique is truly quantitative, especially in organic soils. However, the extraordinary contrast between C:N and C:P ratios in plants versus microbes—a factor of 10 to 200 (figs. 8.9, 8.10)—is consistent with our understanding of plant versus microbial physiology,

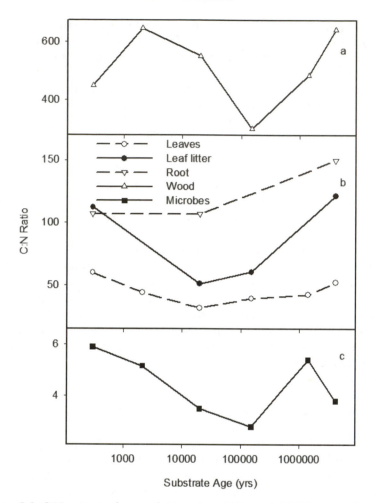

FIGURE 8.9. C:N ratios in the wood (Matzek and Vitousek 2003), roots (Ostertag and Hobbie 1999, Vitousek unpublished), leaves (Vitousek et al. 1995), and leaf litter (Herbert and Fownes 1999) of *Metrosideros polymorpha* and in microbial biomass (Torn et al. submitted) across the Hawaiian age gradient. All plant tissues have much wider C:N ratios than do microbes, particularly in the infertile young and old sites on the gradient.

and with observations in a wide range of ecosystems (Paul and Clark 1996, Elser et al. 2000, Sterner and Elser 2002). Microbial N:P ratios also are low relative to those in most plant tissues (except wood), although to a much lesser extent (fig. 8.11). Relative to their own requirements, microbes utilize C-rich and N- and particularly P-poor resources

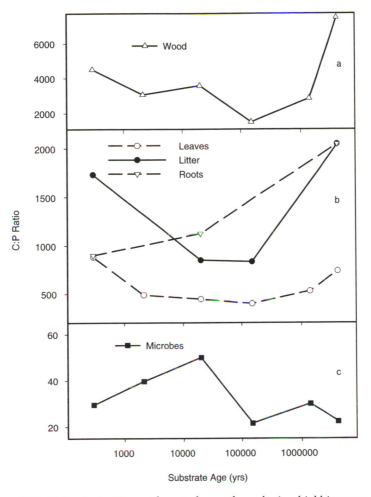

FIGURE 8.10. C:P ratios in *Metrosideros polymorpha* and microbial biomass across the Hawaiian age gradient; sources as in fig. 8.9.

across all of the Hawaiian sites, particularly the youngest and oldest (figs. 8.9, 8.10). Under these conditions, microbial growth efficiency should be low, and the N and P that microbes do acquire should be retained effectively within their cells. Moreover, because P in plant tissues appears even less abundant than N (relative to microbial requirements) across most of the age gradient, a simple stoichiometric analysis would suggest that P should be retained within microbes more effectively than is N in most of the sites.

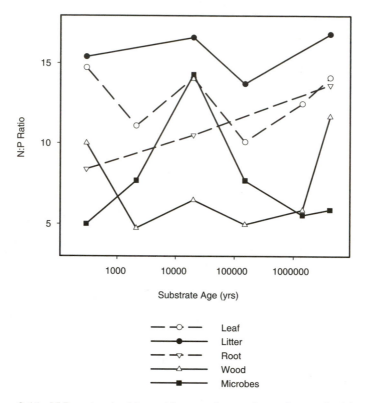

FIGURE 8.11. N:P ratios in *Metrosideros polymorpha* and microbial biomass across the Hawaiian age gradient; sources as in fig. 8.9. Although microbes and wood have relatively low ratios, the variation in N:P ratios among plant tissues and between plants and microbes is much less than that for C:N or C:P ratios.

In fact, the high C:N and C:P ratios in low-nutrient young and old sites are associated with slow rates of decomposition and even slower rates of mineralization in those sites (figs. 4.14, 4.15). However, several features of litter and soil organic matter and their decomposition make it difficult to argue that elemental stoichiometry per se proximately causes slow N and P cycling in infertile Hawaiian forests, and challenge the implication that P should be in short supply relative to N in most of the sites.

First, there is abundant evidence that the C and energy richness of plant litter and soil organic matter (SOM) is more apparent than real, especially within infertile forest ecosystems. Much of organic C is in forms that are recalcitrant to decomposition (initially lignin and polyphenols, ultimately humus and related compounds), and decomposition can be limited by the "quality" of C, rather than by the supply of N or P (Paul

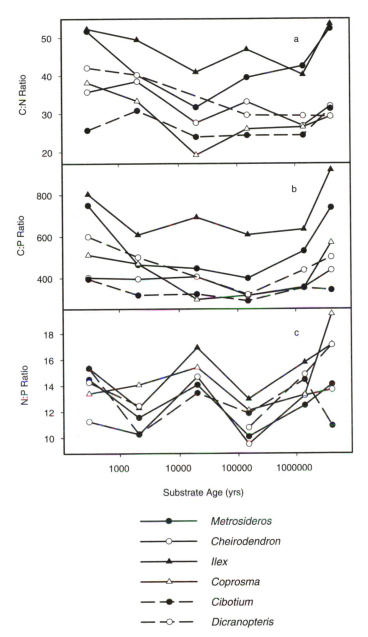

FIGURE 8.12. C:N:P stoichiometry in leaves of six species that occur across the Hawaiian substrate age gradient, from Vitousek et al. (1995b). (a, b) C:N and C:P ratios generally are wider in the infertile young and old sites. (c) N:P ratios vary less among species; they differ consistently between sites but not systematically along the age gradient.

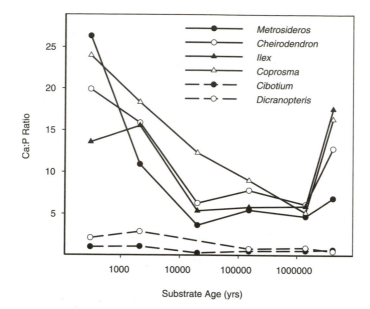

FIGURE 8.13. Ca:P ratios in leaves of the six species in fig. 8.12. Ferns (dashed lines) have much narrower Ca:P ratios than angiosperms in all sites.

and Clark 1996). In Hawai'i, we observe that added N (and P) enhance the decomposition of high quality litter (with low concentrations of lignin and polyphenols), but added nutrients have little or no effect on the decomposition of lower quality litter (fig. 5.9) (Hobbie 2000). Indeed, many recalcitrant compounds and complexes influence decomposition and nutrient cycling persistently enough to be considered factors in stoichiometry, at least on time scales relevant to microbial growth and turnover.

Second, although grazers might gain access to most or all the elements in material they consume, most recalcitrant compounds are too large to be transported across decomposers' cell membranes; rather, they must be processed by extracellular enzymes until their constituents are small enough to be used (Sinsabaugh 1994, Carreiro et al. 2000).

Third, different elements are bound differently within litter and SOM. Most N in litter and SOM bonds directly to C in protein-polyphenol complexes and other large, recalcitrant organic compounds, but most P is present as ester phosphates (held by a C–O–P bond) (McGill and Cole 1981). Extracellular phosphatase enzymes can release inorganic P without breaking down the organic compound or complex that contains it, in SOM if not in litter (Gressel et al. 1996), while the release of N generally requires the coordinated action of several enzymes. In effect, organic compounds themselves must be broken down before inorganic N or soluble amino acids can be released in biologically available forms (McGill and Cole 1981, Hunt et al. 1983).

Fourth, once released, P can be adsorbed by a number of pathways, some of which are strong enough to remove it from circulation more or less permanently (Uehara and Gillman 1981, Sollins et al. 1988).

These features of litter and SOM decomposition allow for substantial flexibility in within-system nutrient cycling, especially for P. Enzymes are built of C and N—and organisms whose growth is constrained by P can invest C and N in the prospect of obtaining P. In contrast, N-limited organisms would need to cast large quantities of N into the environment, in several distinct enzymes, in the prospect of obtaining more N (Schimel and Weintraub 2003). Accordingly, while P mineralization can run ahead of the stoichiometry of SOM decomposition when P is in short supply, that flexibility is much-reduced or absent in the mineralization of N.

We evaluated the flexibility of enzyme-mediated N and P cycling along the Hawaiian age gradient by determining changes in enzyme activity following long-term fertilization (Olander and Vitousek 2000, Treseder and Vitousek 2001a). Adding N to N-limited sites increases productivity and therefore the demand for P and other elements; similarly, adding P to P-limited systems increases demand for N. We found that both decomposers and mycorrhizal roots produced more extracellular phosphatase following N additions (fig. 8.14), but that decomposers did not produce more chitinase (which breaks down an important N-containing compound) following P additions, regardless of whether or not P limited plant production. This greater flexibility of P cycling also influenced nutrient losses, as demonstrated by the retention of P relative to N and C as dissolved organic matter leaches through soils (fig. 7.6) (Hedin et al. 2003).

Inputs and Outputs

Biological stoichiometries and flexibilities affect both element supply and demand, and so influence the cycling of biophilic nutrients within ecosystems. However, a very different set of geochemical stoichiometries and flexibilities affect inputs (and to a lesser extent outputs) of elements, and thus control the nutrient capital of ecosystems on longer time scales. Of the five major pathways of element input to ecosystems across the Hawaiian Islands—atmospheric deposition of marine aerosol, basalt weathering, continental dust, volcanic sources, and biological N fixation (chapter 6)—most supply elements at ratios that are primarily set by the geochemical stoichiometries of rocks and seawater, rather than by the requirements of organisms.

The most consistent of these geochemical stoichiometries is that of seawater, as expressed in the marine aerosol that dominates atmospheric

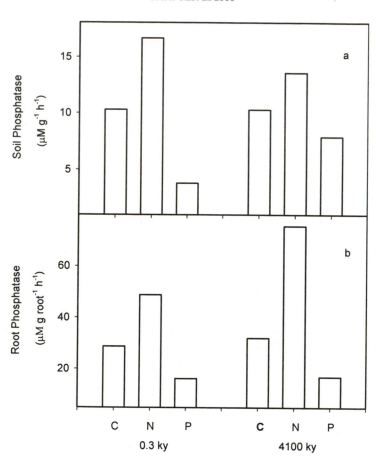

FIGURE 8.14. Phosphatase enzyme activity in control and fertilized plots of the youngest and oldest sites on the substrate age gradient. (a) Phosphatase activity in the soil, from Olander and Vitousek (2000). (b) Root-associated phosphatase activity, from Treseder and Vitousek (2001a). Both root and soil phosphatase activities increase in response to N additions and decrease in response to added P.

deposition in Hawaiʻi (Carrillo et al. 2002). Ratios of elements in seawater are well-defined; a site cannot receive one kg/ha of K from marine aerosol without also receiving 27 kg/ha of Na (table 8.1). By measuring the flux of an element in marine aerosol, and knowing its ratio to other elements in seawater, we can calculate inputs of all the major elements supplied by this pathway.

Elements in basalt also are present at characteristic ratios, but the stoichiometry of basalt weathering is more complex than that of marine

TABLE 8.1

Stoichiometry of the major sources of element inputs to Hawaiian ecosystems, presented as element:P ratios (mass basis). Information from chapter 6, except that ratios in plant tissue are from Bowen (1979).

Pathway	Ca/P	Mg/P	K/P	N/P	S/P	Cl/P	Si/P
Basalt Weathering							
Tholeiitic	106	87	3.6	< 0.1	2.3	< 0.1	346
Alkalic	20	11	3.8	< 0.1	0.6	< 0.1	94
Marine Aerosol	169000	518000	161000	16	1090000	7800000	23
Continental Dust	44	19	41	< 0.1	1.4	0.9	447
Volcanic Inputs	0	0	24	24	622	288	0
Plants	8	4	10	15	5	1	0.4

aerosol. First, Hawaiian basalts differ slightly within and between eruptions (Wright and Helz 1987), and more substantially between the different stages of a volcano's life cycle (Clague and Dalrymple 1987). To calculate element inputs via weathering (Chadwick et al. 1999; chapter 6), we defined two basalt stoichiometries based on an average chemistry for shield-building (tholeiitic) eruptions versus post-shield (alkalic) eruptions (table 8.1). Second, the minerals within basalt contain elements at different ratios, and weather at different rates. Third, some of the elements (e.g., P, Al) are relatively immobile, and weathering inputs of these elements are shifted to later in soil development (fig. 6.3).

These differences can affect the ratios of elements added to ecosystems via basalt weathering substantially—one kg of P entering the 0.3 ky site is accompanied by 85 kg of Ca; at the 4100 ky site the ratio is one kg of P to 15 kg of Ca. At the same time, basalt weathering declines dramatically over time, such that the young site in fact receives 0.51 kg ha^{-1} yr^{-1} of P via weathering, while the oldest site receives only 0.008 kg ha^{-1} yr^{-1} (table 6.1).

Continental dust carries elements to the Hawaiian Islands in a characteristic stoichiometry near that of average element ratios in the upper crust, and quite different from that of Hawaiian basalt (table 8.1; Kurtz et al. 2001). Although dust inputs are small, the decline in basalt weathering over time is so dramatic that at the oldest site on the age sequence, dust supplies the majority of P inputs (Chadwick et al. 1999). Volcanic inputs represent something of a catch-all, including S derived from volcanic

SO_2 emissions, Cl volatilized as lava flows enter the ocean, P volatilized from hot rock, and N that is thermally fixed on flowing lava (Carrillo et al. 2002, Benitez-Nelson personal communication). Empirically, however, volcanic activity provides a significant and reasonably well-characterized suite of element inputs to the youngest sites, and these inputs too are independent of organisms' requirements.

In contrast, biological N fixation can add N—and only N—when and where organisms require it. In this sense, N fixation represents a source of flexibility in element inputs, one that I will turn to shortly.

The summary of input stoichiometries in table 8.1 can be used in several ways. First, these geochemical stoichiometries are quite different from any biological stoichiometries, as illustrated here by Bowen's (1979) summary of the chemistry of angiosperms. N is abundant in plants but minor in most sources of input—and without biological N fixation and/or anthropogenic N sources, N supply clearly should constrain plant growth. Second, compared to biological stoichiometries, P is depleted relative to most other rock-derived elements in all of the major pathways of element input (table 8.1). For example, the ratio of Ca to P in plants averages 8 (Bowen 1979)—but marine aerosol, basalt weathering, and continental dust have Ca:P ratios of > 100,000, 20–105, and 44, respectively. Consequently, it makes sense that among the elements with geochemical sources, P limits the growth of plants and the productivity of ecosystems on old substrates in the Hawaiian Islands (Herbert and Fownes 1995), and over more of Earth than any other rock-derived element. Not even the relative immobility of P or the relative flexibility of within-system P cycling can keep it from limiting plant growth on old soils.

Third, geochemical stoichiometries can be used to evaluate conditions other than those summarized here. For example, how would overall element inputs differ for a site on basalt bedrock in the center of a continent, where marine aerosol might make a contribution 1–10% of that observed in the Hawaiian Islands—and continental dust could contribute 10–1000 times more (Simonson 1995)?

Finally, as long as the stoichiometry of element inputs to ecosystems is controlled geochemically, then inputs of elements that limit ecosystem processes will be accompanied by inputs of non-essential elements, and of other essential elements far in excess of biological requirements. To the extent that microbes specifically weather particular minerals to obtain particular elements, weathering might be more flexible than suggested here (Newman and Banfield 2002, Blum et al. 2002). Still, I think it's fair to say that although elements cycle within ecosystems at ratios established by biological stoichiometries, most of them flow into terrestrial ecosystems at ratios that are controlled primarily by geochemical processes.

Biological N Fixation

Although inputs of most elements are constrained by the stoichiometries of their sources, biological N fixation provides flexibility to N inputs. Organisms that can draw upon the inexhaustible supply of N_2 in the atmosphere should have an advantage over others where N is in short supply—and as a byproduct of their activity, they should increase the supply of fixed N in the system as a whole. When N is abundant, N fixation is repressed—in effect shutting off the process when N is abundant and allowing it to be active when N is in short supply (Hartwig 1998). The quantity of N that can be fixed annually exceeds 100 kg ha^{-1} yr^{-1} in many sites (Boring et al. 1988, Sprent and Sprent 1990), more than enough to overwhelm any plant-soil-microbial feedback and to fill any sink relatively rapidly.

This ability of biological N fixation to respond flexibly to N deficiency formed much of the basis for Redfield's (1958) analysis of marine stoichiometry. More recent analyses have come to similar conclusions, in most freshwater (Schindler 1977), many marine (Tyrrell 1999), and some terrestrial (Eisele et al. 1989, Smith 1992, Crews 1993) ecosystems. However, primary productivity in many temperate and boreal ecosystems, most grasslands, and most estuaries is limited by N, despite the presence of organisms with the capacity to fix N in all these systems (Vitousek and Howarth 1991). What constrains biological N fixation and keeps it from reversing N limitation over much of the surface of Earth?

Answering these questions is complicated by the diversity of N-fixing organisms, which most importantly includes symbiotic associations between plants and prokaryotes, free-living cyanobacteria, and some heterotrophic bacteria. A number of experimental studies, models, and reviews have evaluated constraints to symbiotic and cyanobacterial N fixation in terrestrial and marine ecosystems (Vitousek and Howarth 1991, Hartwig 1998, Rastetter et al. 2001); possible controls that have been identified include:

- Energetic constraints, because the cost of acquiring N via fixation is greater than that of taking up fixed N, especially as NH_4 (Gutschick 1981). This cost could reduce the shade tolerance of plants that are dependent on symbiotic N fixation, thereby keeping them from colonizing closed-canopy, N-limited ecosystems (Vitousek and Field 1999, Rastetter et al. 2001).
- Differences in stoichiometry, in that symbiotic N fixers might require more P than do otherwise-similar non-fixers (though see Sprent 1999). Consequently, a low availability of P—and/or Fe or Mo (both required by the nitrogenase enzyme)—could constrain

rates of N fixation (Cole and Heil 1981, Howarth and Cole 1985, Silvester 1989, Howarth et al. 1999, Karl et al. 2002). Moreover, legumes have an N-rich stoichiometry whether or not they fix N (McKey 1994)—a distinction that probably has a deep evolutionary history, and one they may share with the relatively closely-related actinorrhizal N fixers (Soltis et al. 1995).

• Grazing, because the N-rich stoichiometry of legumes in particular makes them a higher-quality food source for grazers, which have a C:N ratio in their tissues far below that of most terrestrial plants (White 1993, Ritchie et al. 1998, Sterner and Elser 2002).

Each of these mechanisms has been shown to be important in particular situations—and models suggest that all of them have the capacity to suppress the flexibility associated with N fixation, and so contribute to sustaining N limitation (Vitousek et al. 2002). Unfortunately, the Hawaiian age gradient is not a good model system for understanding constraints to symbiotic N fixation. Although one native leguminous tree (*Acacia koa*) is widespread across the islands (Pearson and Vitousek 2001, 2002), it is extremely sparse or absent in the main age gradient sites. However, the gradient is useful for understanding heterotrophic N fixation associated with decomposing litter. The ability to fix N could benefit heterotrophic bacteria because their C:N ratio is extremely narrow relative to that of plant litter (fig. 8.9)—and heterotrophic bacteria with the capacity to fix N are ubiquitous, or nearly so. Nevertheless, rates of heterotrophic N fixation in decomposing litter generally are relatively low in Hawai'i and elsewhere, particularly in infertile forests (fig. 6.6) (Cleveland et al. 1999, Crews et al. 2000).

What constrains rates of heterotrophic N fixation in N-limited ecosystems? To address this question, we made use of the array of litter decomposition measurements and experiments described in chapters 4 and 5, supplemented with an additional experiment (Vitousek and Hobbie 2000) that involved decomposing *Metrosideros* litter that differed widely in chemistry in the control, +P, +T (all essential nutrients other than N and P), and +P +T fertilizer plots of the 0.3 ky site. N fixation was estimated using acetylene reduction calibrated with $^{15}N_2$ fixation. Concentrations of nutrients (N, P, Mo, Ca, K) and recalcitrant C compounds (lignin, polyphenols) varied from 2.5-fold to nearly 10-fold among the sources of *Metrosideros* litter included in the experiment (table 8.2).

Overall, lignin concentrations in litter provided the best predictor of rates of N fixation. Where lignin was < 15%, N equivalent to 30–50% of the initial N in litter was fixed during decomposition, representing a substantial addition each time N cycles through litter (Vitousek and Hobbie 2000). In contrast, where lignin was > 18% of litter dry mass, integrated

TABLE 8.2
The range in chemical properties of *Metrosideros polymorpha* leaf litter
used to evaluate controls of heterotrophic N fixation. From Hobbie (2000),
Hobbie and Vitousek (2000), and Vitousek and Hobbie (2000).

Constituent	Low	High
Nitrogen (%)	0.19	0.90
Phosphorus (%)	0.013	0.27
Lignin (%)	10.7	28.5
Calcium (%)	0.61	2.58
Magnesium (%)	0.12	0.58
Potassium (%)	0.04	0.15
Molybdenum (mg/kg)[1]	0.23	3.0

Notes
[1]Mo was determined only on a subset of litter types; the actual range in concentrations
could be wider.

N fixation amounted to only 3–8% of initial N in the litter. The effects of
P, Mo, and other nutrients on N fixation were small, variable, and wholly
explicable in terms of C chemistry. Although litter produced in P-fertilized plots had significantly greater rates of N fixation than litter from
control plots (a 1.7-fold increase), it also had significantly lower lignin
concentrations, enough to explain a 1.6-fold increase in N fixation (Vitousek and Hobbie 2000).

Why this negative association between recalcitrant C compounds and
heterotrophic N fixation? As discussed in chapter 5, Hobbie (2000) demonstrated that where lignin concentrations are low, additions of N strongly
enhance rates of decomposition. However, where lignin concentrations are
high, added N has little or no effect on decomposition (fig. 5.9, repeated
as fig. 8.15a). This result makes sense if C quality, more than nutrient
availability, controls rates of decomposition in infertile sites. Under these
circumstances, there would be no advantage to a heterotroph that invested
energy in fixing N, because neither decomposition nor the growth of decomposer populations is limited by N supply where C quality is low. Where
C quality is high (low lignin), N supply can limit rates of decomposition—
and there can be a substantial benefit to heterotrophic N fixation, just as
we observed (fig. 8.15b).

Overall, N limitation to forest growth causes trees to produce (or
selects for genotypes that produce) long-lived leaves rich in C, structure,

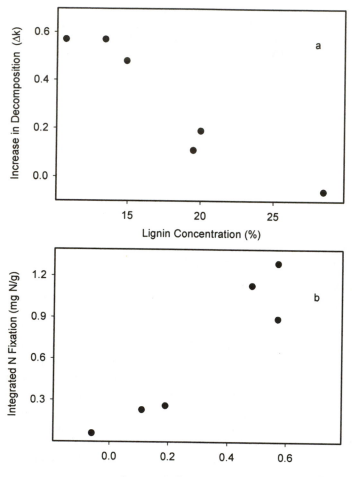

FIGURE 8.15. Decomposition and heterotrophic N fixation in relation to C quality (lignin concentration) and N limitation to decomposition for *Metrosideros polymorpha* leaf litter, from Hobbie (2000), Hobbie and Vitousek (2000), and Vitousek and Hobbie (2000). (a) N additions stimulate decomposition of low-lignin litter, but have little or no effect on high-lignin litter (also fig. 5.9). (b) Heterotrophic N fixation is greater in low-lignin litter types in which N supply limits decomposition; in contrast, decomposition of high-lignin litter is not N-limited and does not support substantial N fixation.

and defense. When these leaves senesce, their decomposition is limited by C quality, not by N supply, and heterotrophic bacteria gain little or no benefit from fixing N. This disconnect between plants and decomposers has the effect of constraining heterotrophic N fixation in ecosystems where N limits production, and so of maintaining N limitation to production (Vitousek and Hobbie 2000).

DIFFERENCES IN POPULATIONS, SPECIES, AND DIVERSITY

As discussed in chapter 2, the few lineages that naturally found their way to Hawai'i generally have radiated to occupy a much broader range of environments than do most continental species—a feature that allows us to hold species composition relatively constant across an extraordinarily broad range of conditions. Here, I ask two questions: Could relatively subtle differences in biota across the Hawaiian age gradient affect the functioning of ecosystems there? Would the functioning of Hawaiian ecosystems across the gradient differ if biological diversity in Hawai'i were substantially greater?

Biological Differences and Ecosystem Functioning

POPULATION–LEVEL DIFFERENCES IN *METROSIDEROS POLYMORPHA*

For *Metrosideros*, the specific epithet "*polymorpha*" accurately implies a highly variable species. Although molecular phylogenies suggest that its dispersal across the Pacific is relatively recent (Wright et al. 2000, 2001), morphological varieties of *Metrosideros* have differentiated substantially along environmental gradients in Hawai'i (Stemmermann 1983, Dawson and Stemmermann 1990), and common-garden studies demonstrate that many of these differences are genetically based (Cordell et al. 1998). Moreover, moderate differences in isozyme profiles of different *Metrosideros* populations have been documented (Aradhya et al. 1991, 1993). Because *Metrosideros* accounts for most of the biomass along the Hawaiian age gradient, its widespread distribution offers an unusual opportunity to explore the ecosystem-level consequences of population-level variation within a species (Holland et al 1992). To this end, Kathleen Treseder established a common garden of *Metrosideros* from the 0.3, 20, and 4100 ky sites, and determined their isozyme profiles, plant tissue chemistry, growth rates, and responsiveness to nutrient additions under identical conditions. Isozyme analysis yielded moderate differences among age-gradient populations, with those from the 0.3 ky site being the most distinct (Treseder and Vitousek 2001b).

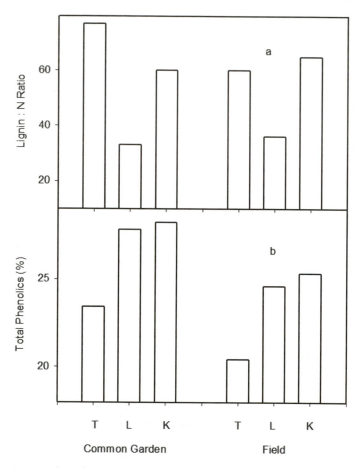

FIGURE 8.16. Foliar chemistry of *Metrosideros polymorpha* grown under identical conditions in a common garden, and of field-grown trees (Treseder and Vitousek 2001b). (a) The population from the fertile 20 ky Laupāhoehoe site produces litter with a low lignin:N ratio in both garden and field, consistent with its relatively rapid rate of decomposition. (b) The population from the 0.3 ky Thurston site produces leaves with lower concentrations of soluble polyphenols in both common garden and field.

Several characteristics of the common-garden plants differed among populations in ways consistent with differences observed in the field. Senescent leaves of common-garden plants from the 20 ky site had narrower lignin-N ratios than those from the younger and older sites (fig. 8.16a)—just as we observed in the field, where decomposition of litter was indeed more rapid in the fertile, intermediate-aged site (fig. 4.14). Although field measurement could not distinguish any influence of plant genotype from the substantial differences in nutrient availability among sites, common-garden results suggest that population-level differences could contribute to the pattern in the field. Concentrations of soluble polyphenols in leaves also varied among common garden plants in the same pattern as in the field (fig. 8.16b) (Hättenschwiler et al. 2003). Finally, common-garden plants from the P-limited 4100 ky site accumulated excess P to much higher levels than did plants from other sites—just as occurs in P-fertilized field plots there (Treseder and Vitousek 2001b, chapter 4). In a separate study, common-garden populations of *Acacia koa* collected across a drier substrate age gradient behaved similarly, with plants from 4100 ky substrates accumulating much more P when fertilized than those from younger sites (Pearson 1998).

Some of the responses of the fertilization experiments reinforce the suggestion that genetic differences among *Metrosideros* populations influence aspects of ecosystem functioning. Regardless of how long they were fertilized, trees in the N-limited 0.3 ky site never approached the foliar or litter N concentrations observed in the more fertile intermediate-aged sites (chapter 5). I believe that this lack of flexibility exhibited by the *Metrosideros* population that dominates the 0.3 ky site contributes to the relatively slow decomposition of leaf litter there, even following long-term fertilization (figs. 4.14, 5.2, 5.6). Given enough time, a more nutrient-rich and -responsive variety of *Metrosideros* might come to dominate N-fertilized plots (Stemmermann 1983), and decomposition/nutrient cycling could then change substantially. However, the biotic inertia represented by this long-lived, relatively unresponsive population could delay the full expression of any plant-soil-microbial positive feedback, for decades and perhaps even centuries.

The likely importance of population-level changes in response to changes in nutrient availability requires that I reconsider the interacting time scales discussed at the beginning of this chapter, particularly those relating to plant-soil feedbacks (fig. 8.2), because the replacement of an unresponsive variety by one better suited to high-nutrient conditions adds a new time scale to that analysis. I modified the Explore model to incorporate population replacement by introducing a delay between the imposition of a change in nutrient availability and the realization of a change in plant nutrient use efficiency and decomposability. The earlier analyses (fig. 8.2) assumed no delay, corresponding to a phenotypic response to altered nutrient availability. Where the dominant population occupying a site must be replaced before the feedback can be expressed, however, the model response depends on the delay before the more responsive variety dominates the site versus the time required for nutrient inputs to accumulate and reverse the effects of a decrease in the nutrient availability

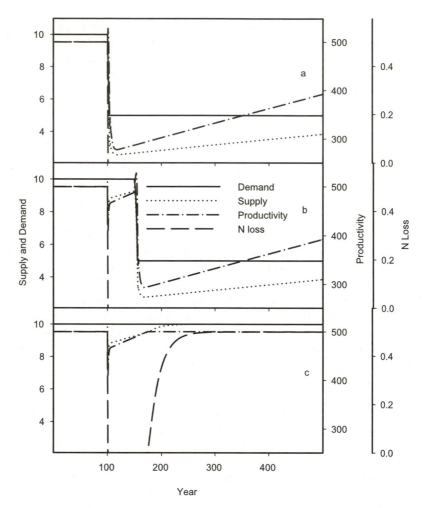

FIGURE 8.17. The consequences of introducing a lag between the initiation of supply-demand imbalance and the development of plant-soil feedback within the Explore model. I assume that a lag occurs because the plant population occupying a site is unresponsive to a change in nutrient availability, and the feedback cannot occur until the original population is replaced by a more responsive population (or species). (a) No delay between the introduction of supply-demand imbalance (by adding 50 units of C to soil each year) in year 100 and the development of a feedback; these conditions are identical to those in fig. 8.2. (b) A 50-year lag in the development of the feedback; supply, demand, productivity, and N losses all decline similarly to fig. 8.17a, but after a 50-year delay. (c) 100-year lag in the development of the feedback; the system accumulates enough N during the lag period that the feedback never gets going.

(fig. 8.17). Under the conditions simulated here, a 50-year lag between the imposition of a decrease in nutrient availability and the replacement of the unresponsive population delays the start of the feedback without altering its nature, but a 100-year lag allows nutrient inputs to accumulate to the point that no plant-soil-microbial feedback develops (fig. 8.17b,c). While the model is exploratory, I believe that the implications of these lags and thresholds are worth evaluating further.

UNDERSTORY PLANTS

Although *Metrosideros* dominates and several other species occur across the complete Hawaiian age gradient, there are substantial differences in understory species composition and diversity (Kitayama and Mueller-Dombois 1995). The most striking of these is a progressive decline in *Cibotium* tree ferns, which dominate the subcanopy in young sites but decline and virtually disappear by the oldest site (fig. 3.12). I don't know how this change influences the functioning of ecosystems, but given the very different cation stoichiometry and carbon chemistry of ferns compared to angiosperms (fig. 8.13) and the slow decomposition of tree fern litter (Scowcroft 1997, Allison and Vitousek submitted), it is a question worth exploring.

SOIL ORGANISMS

Just as nutrient use efficiency and litter decomposability can be influenced by plant genotypes (and species), decomposition and nutrient mineralization could be shaped by differences in the community of soil organisms (Wardle 2002). We found that a common litter type decomposed more rapidly in fertile intermediate aged sites than in infertile young and old sites—even when control plots in the fertile site are compared with long-term fertilized plots in the infertile ones (figs. 4.14, 5.6, table 4.3) (Crews et al. 1995, Hobbie and Vitousek 2000). One plausible explanation is that fertile soils support biological communities that promote rapid decomposition, and even > 10 years fertilization does not reproduce this effect because like tree populations, communities of soil organisms respond slowly to changes in nutrient availability. If so, either differences in soil microorganisms or soil fauna—or both—could cause differences in decomposition among the sites.

For microorganisms, the relatively recent development of culture-independent biochemical and genetic techniques has provided access to a surprisingly diverse array of organisms (Tiedje et al. 1999). Research along the Hawaiian age gradient contributed to the development of these techniques (Nusslein and Tiedje 1998), and continuing research has documented differences in microbial communities (Balser 2002) without as yet linking differences in community composition to differences in ecosystem function.

The influence of soil fauna on rates of decomposition and other eco-logical processes has been studied far longer than that of microbial com-munities (Darwin 1881), and there is good evidence that soil fauna affect rates of decomposition and nutrient turnover in tropical ecosystems in particular (Heneghan et al. 1999, Cuevas 2001). David Foote found soil mesofauna to be more abundant in the relatively fertile intermediate-aged sites along the Hawaiian age gradient (D. Foote personal communication). This pattern, coupled with the observation that site fertility has a stronger effect on the decomposition of a common substrate when the litter is de-composed in coarse-mesh litterbags that allow faunal access rather than fine-mesh litterbags that prevent it (Ostertag and Hobbie 1999), suggests that differences in soil faunal communities contribute to differences in decomposition/nutrient regeneration across the age gradient.

BIOLOGICAL INVASIONS

The extensive introduction of species from outside Hawai'i represents the most serious threat to the maintenance of native populations, species, and ecosystems—and at the same time, an opportunity to evaluate how changes in species composition affect the functioning of ecosystems. I will focus on plant invasions here, because oceanic islands (including Hawai'i) were notably depauperate in many groups of animals, including mammals, ants, and earthworms (Carlquist 1980, Roughgarden 1995); consequently, in-troduced animals often represent an ongoing disturbance rather than a change in species composition (Cuddihy and Stone 1990, Cole et al. 1992).

The best-documented plant invader on or near the substrate age gradient is the actinorrhizal N fixer *Myrica faya*, which invades young, N-limited sites in Hawai'i Volcanoes National Park—including the 0.3 ky site (where we are actively excluding it) and younger sites nearby. The productivity of these ecosystems is limited by N (fig. 5.1) (Vitousek et al. 1987), and *Myrica* increases the inputs of N substantially, from about 11 kg N ha^{-1} yr^{-1} without *Myrica* (fig. 6.12a) to about 30 kg N ha^{-1} yr^{-1} with it (Vi-tousek and Walker 1989). Some of the N fixed by *Myrica* becomes available to other organisms relatively rapidly (Vitousek and Walker 1989, Matson 1990), and can affect the growth of colonizing plants (Walker and Vitousek 1991) and even earthworm populations (Aplet 1990; fig. 8.18). The clear ecosystem-level changes caused by this invasion demonstrate that the ab-sence of a colonizing N-fixer from the native Hawaiian flora strongly affects the functioning of Hawaiian ecosystems on young substrates.

Later in ecosystem development, *Metrosideros* trees growing on fertile intermediate-aged sites have higher tissue nutrient concentrations, more rapid leaf turnover, and more rapid rates of litter decomposition and nu-trient cycling than those growing on infertile sites (figs. 4.6, 4.14, 4.15). Could an invading species that is adapted to fertile soils in its native

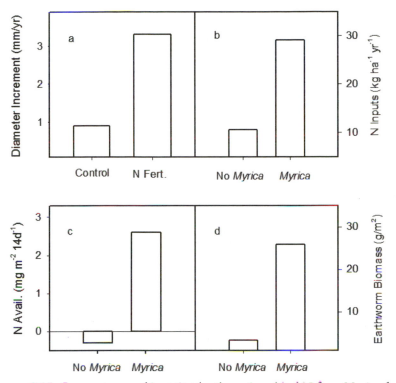

FIGURE 8.18. Consequences of invasion by the actinorrhizal N fixer *Myrica faya* near the 0.3 ky site on the Hawaiian age gradient, from Vitousek and Walker (1989) and Aplet (1990). (a) In the absence of *Myrica*, the growth of *Metrosideros polymorpha* trees is limited by N. (b) *Myrica* invasion increases overall N inputs substantially; these values are revised from Vitousek and Walker (1989) to reflect calculated N inputs from chapter 6. (c) N fixed by *Myrica* is available to other organisms, as the increase in N mineralization under *Myrica* versus *Metrosideros* demonstrates. (d) *Myrica* affects populations of other organisms, as illustrated by earthworm biomass under *Myrica* versus *Metrosideros*.

range display the same properties to an even greater extent, and so drive a stronger plant-soil-microbial positive feedback than does *Metrosideros*? David Rothstein evaluated the consequences of invasion by the Mexican tree *Fraxinus uhdei*, which was planted 60+ years ago near the 20 ky site. *Fraxinus* produces thinner, higher-nutrient litter with lower concentrations of recalcitrant C compounds than does *Metrosideros* in comparably rich sites (table 8.3). *Fraxinus* also grows larger, casts deeper shade (as is readily detectable via satellite imagery as well as from below the canopy), and supports higher productivity. Finally, *Fraxinus* litter decomposes more

TABLE 8.3

Litter properties of introduced *Fraxinus uhdei* and native *Metrosideros polymorpha* collected in the Laupāhoehoe Forest Reserve near the 20 ky site on the age gradient. From Rothstein et al., in press.

	Metrosideros	Fraxinus
Leaf Mass/Area (g/m^2)	137	75
N (%)	0.81	1.15
P (%)	0.053	0.101
Lignin (%)	18.9	15.1
Soluble polyphenols (%)	9.2	6.3
Tannin (%)	6.4	0.4

rapidly and regenerates available N and P much more quickly than does *Metrosideros* (fig. 8.19) (Rothstein et al. in press), suggesting that *Fraxinus* could drive a positive feedback towards rapid nutrient cycling in nutrient-rich sites even more effectively than does the *Metrosideros* population adapted to rich sites. The time-scale of species replacement during invasion—or in response to a change in nutrient availability—could thus be important in the same way as is population replacement (fig. 8.17), but most likely with far greater effects.

Diversity and Ecosystem Functioning

The previous section asked if existing biological differences across the Hawaiian age gradient, though relatively small, could nevertheless influence ecosystem properties. Here, I turn the question around, asking how ecosystem properties along the gradient might differ if biological differences among the sites were much larger. To make my approach more concrete—*Metrosideros polymorpha* is an amazing tree, with the ability to dominate an extraordinarily broad range of sites. However, to do so it must be something of a generalist; it can't do all things equally well. Even though it has radiated into genetically distinct populations, it still represents a single, relatively recent lineage, and gene flow among its populations probably constrains its further differentiation substantially. I speculate that the ability of *Metrosideros* to acquire and utilize resources in most of the places it dominates should differ from what more narrowly distributed species that specialized in particular environmental conditions could achieve. If so, ecosystems dominated by those specialists could function differ-

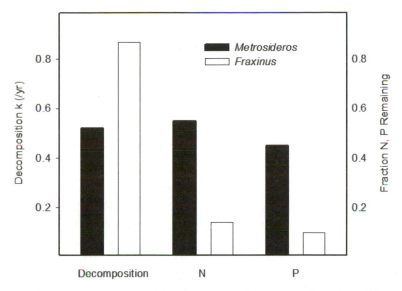

FIGURE 8.19. Litter decomposition and rates of N and P release in leaf litter of *Metrosideros polymorpha* and the introduced tree *Fraxinus uhdei* (tropical ash) in a relatively nutrient-rich area near the 20 ky site on the Hawaiian age gradient, from Rothstein et al. (in press). Tropical ash decomposes more rapidly and releases N and P into available forms much more rapidly than does *Metrosideros*, potentially enhancing the plant-soil-microbial feedback in relatively nutrient-rich sites.

ently (perhaps being more productive and/or drawing nutrients down to lower minimum concentrations) than comparable *Metrosideros*-dominated ecosystems.

This suggestion represents a real-life application of the much-discussed "sampling effect" that crops up in experimental studies of biological diversity and ecosystem functioning (Wardle 1999, Tilman et al. 2001, Hooper et al. in press). Due to the vagaries of species selection, a more diverse experimental plot is more likely to contain one or more highly productive (or nutrient retaining, or drought resistant, etc.) species than is a less diverse plot. Similarly, a diverse flora would be likely to include at least one species that could do "better" than *Metrosideros* in young N-limited sites, at least another that would do better in relatively fertile intermediate-aged sites, and so on across nearly all of the broad range of sites that *Metrosideros* dominates.

Alternatively, ecosystems in regions with a more diverse flora could function differently due to a greater diversity of species within sites, because members of an array of species could have complementary patterns of resource use (e.g., different stoichiometries) or of responsiveness to disturbance and/or environmental change (Tilman et al. 1996, 2001; Hooper and Vitousek 1998). In either case, because the continental tropics have much greater α- and β-diversity than does Hawai'i, ecosystem functioning could differ systematically on continents versus islands.

I see two lines of evidence that could be used to test this suggestion. First, do biological invasions alter the functioning of Hawaiian ecosystems to a greater extent than they alter more diverse continental ecosystems? Second, how do the patterns and magnitudes of variation in productivity and other ecosystem processes in Hawaiian ecosystems compare with those of more diverse continental ecosystems?

BIOLOGICAL INVASIONS

As discussed above, islands are notoriously vulnerable to biological invasions—and Hawaiian ecosystems in particular can be altered substantially by invasion, on the age gradient (figs. 8.18, 8.19) and elsewhere (Hughes et al. 1991, D'Antonio et al. 2001, Mack et al. 2001). Of course, biological invasions also alter the functioning of continental ecosystems (Le Maitre et al. 1995, Evans et al. 2001), although the extent of change generally is believed to be greater on oceanic islands (D'Antonio and Dudley 1995, Denslow 2003). Are invaders of islands successful because they generally use resources more effectively than do island natives—to a greater extent than do invaders versus natives of continents? The answer might well be yes, but I don't think there is sufficient evidence to demonstrate it—and there are other plausible explanations for the relative success of invaders on islands (escape from natural enemies, the ability to tolerate novel disturbance regimes, etc.).

PATTERNS AND MAGNITUDES OF VARIATION IN ECOSYSTEM PROCESSES

The primary productivity of Hawaiian forests generally varies along environmental gradients in the same patterns as tropical forests elsewhere. This statement applies to gradients in elevation and temperature, where productivity decreases and soil C storage increases with decreasing temperature (Jenny et al. 1949; Kitayama and Mueller-Dombois 1994; Raich et al. 1997, 2000; Kitayama and Aiba 2002); gradients in precipitation, where productivity increases with increasing rainfall up to a point, then decreases in wetter sites (Austin and Vitousek 1998, 2000; Clark et al. 2001; Schuur et al. 2001; Austin 2002); and to gradients in ecosystem development, where productivity generally is low in young and old in comparison to intermediate-aged sites (Chapin et al. 1994, Herbert and Fownes 1999). Where information is available, nutrient cycling also varies similarly along gradients in continental and Hawaiian ecosystems. N and P become relatively less available with decreasing temperature (Marrs et al. 1988, Raich et al. 1997, Tanner et al. 1998, Kitayama et al. 2000); the N cycle is progressively more open in drier sites, as indicated by increased ^{15}N enrichment in plants and soils (fig. 3.5) (Austin and Vitousek 1998, Handley et al. 1999); and P availability declines in very old soils (Walker and Syers 1976, Crews et al. 1995, Brenner et al. 2001).

One exception to these patterns is that many (not all) soil developmental sequences in more diverse regions include an early stage dominated by symbiotic N-fixing plants, such as *Alnus* at Glacier Bay and *Casuarina* on Krakatau (Walker and Syers 1976, Chapin et al. 1994, Schlesinger et al. 1998). Where N fixers dominate early in soil development, they increase N availability and cycling much more rapidly than occurs on the Hawaiian age gradient, and thus hasten the transition away from N limitation.

It is more difficult to determine how absolute rates of primary productivity (or other ecosystem processes) in Hawai'i compare with those of continental ecosystems in identical environments—in part because it is difficult to be sure that all else is equal in any such comparison. Annual litterfall in Hawaiian montane forests is less than that in most continents and continental islands (Tanner et al. 1998), although litterfall declines with increasing elevation in all tropical systems. Similarly, aboveground NPP declines

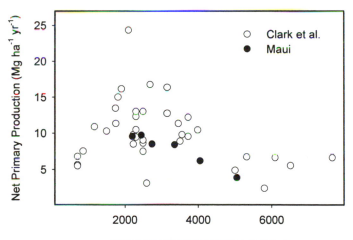

FIGURE 8.20. Tropical forest productivity as a function of mean annual precipitation for a range of mostly continental forests (Clark et al. 2001) (hollow symbols) and for Hawaiian forests arrayed along a precipitation gradient on Maui (dark symbols). Modified from Schuur (2003). Productivity in the Hawaiian sites varies similarly as a function of rainfall as do all of the other sites; in magnitude, the Hawaiian sites fall at the lower end of the range in productivity.

along the Maui rainfall gradient studied by Ted Schuur—with the same pattern as in continental sites (from Clark et al. 2001), but with 20–40% lower productivity across the entire range of rainfall (fig. 8.20) (Schuur 2003). However, the Hawaiian forests in this comparison occupy cooler sites than do most of the continental forests, which could explain much of the difference observed (Schuur 2003). Nevertheless, I believe that a preponderance of evidence points towards a slightly lower productivity in low-diversity Hawaiian forests than in more diverse, climatically-similar tropical forests elsewhere.

THREE FINAL POINTS

I began this book by describing Hawai'i as a model system, and showing how the relative simplicity of the factors that regulate ecosystem structure and functioning there enhances our ability to understand fundamental properties and processes of ecosystems. From that simple beginning, I have spent much of book showing that many of these controls are more complicated than they appear—for example, that the organism factor is complicated by population-level variation within the dominant tree species, and by biological invasions; that both the parent material and time factors are complicated by inputs of continental dust from central Asia; and that the climate factor is complicated by glacial/interglacial climate history. Rather than reducing the value of Hawai'i as a model system, these complications illustrate its value. Identifying them and understanding their importance—in a real model system, not a constructed microcosm from which they would have been excluded, or a complex continental system in which their influence would have been difficult to isolate—has motivated much of our most interesting research.

Second, I believe that continued research in Hawai'i can teach us a great deal more about how the world works. Many of the most exciting research questions and opportunities can be found at interfaces among disciplines, and the Hawaiian Islands are a great model system for understanding aspects of evolution and phylogeny, population ecology, conservation biology, and anthropology/archeology as well as ecosystem ecology/biogeochemistry. I believe that the advantages of Hawaiian research for analyzing each of these fields separately will be multiplied when we consider them interactively.

Finally, the gradients and matrices of ecosystems where our research has been focused are both valuable and beautiful. I hope that I have made their scientific value clear. Part of their beauty is readily apparent, too, in their forests, lava flows, canyons, flowers, birds, and much more—and part of it resides in what they can show all of us (not just researchers) about how the world works. However, despite their value and beauty, not one of these gradients has been conserved or managed as a whole for its value as an ecosystem. Portions of them have been developed with little environmental concern—because those portions contained no endangered species. I understand that a lack of recognition of the value and beauty of ecosystem-level diversity and dynamics is not unique to Hawai'i. I believe that as ecosystem scientists, we need to become more effective advocates for the recognition and conservation of ecosystems.

REFERENCES

Aber, J. D., and C. T. Driscoll. 1997. Effects of land use, climate variation, and N deposition on N cycling and C storage in northern hardwood forests. *Global Biogeochemical Cycles* 11: 639–648.

Aber, J. D., and J. M. Melillo. 2001. *Terrestrial Ecosystems, Second Edition*. San Diego: Harcourt Academic Press.

Aber, J. D., K. J. Nadelhoffer, P. Steudler, and J. M. Melillo. 1989. Nitrogen saturation in northern forest ecosystems. *Bioscience* 39: 378–386.

Aber, J. D., W. H. McDowell, K. Nadelhoffer, A. Magill, G. Berntson, M. Kamakea, S. McNulty, W. Currie, L. Rustad, and I. Fernandez. 1998. Nitrogen saturation in temperate forest ecosystems. *BioScience* 48: 921–934.

Aerts, R. 1996. Nutrient resorption from senescing leaves of perennials: are there general patterns? *Journal of Ecology* 84: 597–608.

Aerts, R., and F. S. Chapin. 2000. The mineral nutrition of wild plants revisited: A reevaluation of processes and patterns. *Advances in Ecological Research* 30: 1–67.

Ågren, G. I., and E. Bosatta. 1988. Nitrogen saturation of terrestrial ecosystems. *Environmental Pollution* 54: 185–197.

———. 1996. *Theoretical Ecosystem Ecology: Understanding Nutrient Cycles*. Cambridge: Cambridge University Press.

Akashi, Y., and D. Mueller-Dombois. 1995. A landscape perspective of the Hawaiian rainforest dieback. *Journal of Vegetation Science* 6: 449–464.

Allison, F. E. 1955. The enigma of soil nitrogen balance sheets. *Advances in Agronomy* 7: 213–250.

Allison, S. D., and P. M. Vitousek. Rapid decomposition and nutrient cycling in invasive Hawaiian understory plant litter. Submitted.

Anderson, W. B., and G. A. Polis. 1999. Nutrient fluxes from water to land: seabirds affect plant nutrient status on the Gulf of California Islands. *Oecologia* 118: 324–333.

Aplet, G. H. 1990. Alteration of earthworm community biomass by the alien *Myrica faya* in Hawai'i. *Oecologia* 82: 414–416.

Aplet, G. H., and P. M. Vitousek. 1994. An age-elevation matrix analysis of Hawaiian rainforest succession. *Journal of Ecology* 82: 137–147.

Aplet, G. H., R. F. Hughes, and P. M. Vitousek. 1998. Ecosystem development on Hawaiian lava flows: biomass and species composition. *Journal of Vegetation Science* 9: 17–26.

Aradhya, K. M., D. Mueller-Dombois, and T. Ranker. 1991. Genetic evidence for recent and incipient speciation in the evolution of Hawaiian *Metrosideros* (Myrtaceae). *Heredity* 67: 129–138.

———. 1993. Genetic structure and differentiation in *Metrosideros polymorpha* along altitudinal gradients in Maui. *Genetics Research* 61: 159–170.

Asbury, C. E., W. H. McDowell, R. Trinidad-Pizarro, and S. Berrios. 1994. Solute deposition from cloudwater to the canopy of a Puerto Rican montane forest. *Atmospheric Environment* 28: 1773–1780.

Atkinson, I.A.E. 1970. Successional trends in the coastal and lowland forest of Mauna Loa and Kilauea Volcanoes, Hawaii. *Pacific Science* 24: 387–400.

Austin, A. T. 2002. Differential effects of precipitation on production and decomposition along a rainfall gradient in Hawai'i. *Ecology* 83: 328–338.

Austin, A. T., and P. M. Vitousek. 1998. Nutrient dynamics on a precipitation gradient in Hawai'i. *Oecologia* 113: 519–529.

———. 2000. Precipitation, decomposition and decomposability of *Metrosideros polymorpha* in native forests on Hawai'i. *Journal of Ecology* 88: 129–138.

Baldwin, B. G. 1997. Adaptive radiation of the Hawaiian silversword alliance: Congruence and conflict of phylogenetic evidence from molecular and non-molecular investigations. In *Molecular Evolution and Adaptive Radiation,* eds. T. J. Givnish and K. J. Sytsma, 103–128. Cambridge: Cambridge University Press.

Baldwin, B. G., and M. J. Sanderson. 1998. Age and rate of diversification of the Hawaiian silversword alliance (Compositae). *Proceedings of the National Academy of Sciences* 95: 9402–9406.

Balser, T. C. 2002. The impact of long-term nitrogen addition on microbial community composition in three Hawaiian forest soils. In *Optimizing Nitrogen Management in Food and Energy Production and Environmental Protection,* eds. J. Galloway, E. Cowling, J. W. Erisman, J. Wisniewski, and C. Jordan, 500–504. Lisse, the Netherlands: Swets and Zeitlinger B. V.

Benitez-Nelson, C. R., S. M. Vink, J. H. Carrillo, and B. J. Huebert. 2003. Volcanically influenced Fe and Al cloud water deposition to Hawaii. *Atmospheric Environment* 37: 535–544.

Bennett, E. M., S. R. Carpenter, and N. F. Caraco. 2001. Human impact on erodable phosphorus and eutrophication: A global perspective. *BioScience* 51: 227–234.

Berendse, F., and R. Aerts. 1987. Nitrogen-use efficiency: A biologically meaningful definition? *Functional Ecology* 1: 293–296.

Binkley, D., and C. Giardina. 1998. Why do tree species affect soils? The warp and woof of tree-soil interactions. *Biogeochemistry* 42: 89–106.

Binkley, D., P. Sollins, R. Bell, D. Sachs, and D. Myrold. 1992. Biogeochemistry of adjacent conifer and alder-conifer stands. *Ecology* 73: 2022–2033.

Binkley, D., Y. Son, and D. W. Valentine. 2000. Do forests receive occult inputs of nitrogen? *Ecosystems* 3: 321–331.

Blum, J. D., A. Klaue, C. A. Nezat, C. T. Driscoll, C. E. Johnson, T. G. Siccama, C. Eagar, T. J. Fahey, and G. E. Likens. 2002. Mycorrhizal weathering of apatite as an important calcium source in base-poor forest ecosystems. *Nature* 417: 729–731.

Bohannan, B.J.M., and R. E. Lenski. 1997. Effect of resource enrichment on a chemostat community of bacteria and bacteriophage. *Ecology* 78: 2303–2315.

Bohlen P. J., P. M. Groffman, T. J. Fahey and M. C. Fisk. Ecosystem consequences of exotic earthworm invasion of northern forests. *Ecosystems,* in press.

Bolin, B., and R. B. Cook (eds.) 1983. *The Major Biogeochemical Cycles and Their Interactions.* New York: John Wiley and Sons.

Boring, L. R., W. T. Swank, J. B. Waide, and G. S. Henderson. 1988. Sources, fates, and impacts of nitrogen inputs to terrestrial ecosystem: review and synthesis. *Biogeochemistry* 6: 119–159.

Bormann, B. T., F. H. Bormann, W. B. Bowden, R. S. Pierce, S. P. Hamburg, D. Wang, M. C. Snyder, C. Y. Li, and R. C. Ingersoll. 1993. Rapid N_2 fixation in pines, alder, and locust: evidence from the sandbox ecosystem study. *Ecology* 74: 581–598.

Bormann, F. H., and G. E. Likens. 1979. *Pattern and Processes in a Forested Ecosystem.* New York: Springer-Verlag.

Bowden, R. D. 1991. Inputs, outputs, and accumulation of nitrogen in an early successional moss (*Polytrichum*) ecosystem. *Ecological Monographs* 61: 207–223.

Bowen, H.J.M. 1979. *Environmental Chemistry of the Elements.* London: Academic Press.

Brenner, D. L., R. Amundson, W. T. Baisden, C. Kendall, and J. Harden. 2001. Soil N and ^{15}N variation with time in a California annual grassland ecosystem. *Geochimica et Cosmochimica Acta* 65: 4171–4186.

Bridgham, S. D., J. Pastor, C. A. McClaugherty, and C. J. Richardson. 1995. Nutrient-use efficiency: a litterfall index, a model, and a test along a nutrient availability gradient in North Carolina peatlands. *American Naturalist* 145: 1–21.

Brimhall, G. H., O. A. Chadwick, C. J. Lewis, W. Compston, I. S. Williams, K. J. Danti, W. E. Dietrich, M. E. Power, D. M. Hendricks, and J. Bratt. 1992. Deformational mass transport and invasive processes in soil evolution. *Science* 255: 695–702.

Bruijnzeel, L. A., and E. J. Veneklaas. 1998. Climatic conditions and tropical montane forest productivity: the fog has not lifted yet. *Ecology* 79: 3–9.

Burke, I. C., W. K. Lauenroth, and W. J. Parton. 1997. Regional and temporal variation in net primary production and nitrogen mineralization in grasslands. *Ecology* 78: 1330–1340.

Burney, D. A., H. F. James, L. P. Burney, S. L. Olson, W. Kikuchi, W. L. Wagner, M. Burney, D. McCloskey, D. Kikuchi, F. V. Grady, R. Gage II, and R. Nishek. 2001. Fossil evidence for a diverse biota from Kaua'i and its transformation since human arrival. *Ecological Monographs* 71: 615–641.

C. *elegans* Sequencing Consortium. 1998. Genome sequence of the nematode C. *elegans:* A platform for investigating biology. *Science* 282: 2012–2018.

Carlquist, S. 1974. *Island Biology.* New York: Columbia University Press.

————. 1980. *Hawaii: A Natural History.* Lawai, Hawaii: Pacific Tropical Botanical Garden.

————. 1982. The first arrivals. *Natural History* 91(12): 20–30.

Carpenter, S. R. 1996. Microcosm experiments have limited relevance for community and ecosystem ecology. *Ecology* 77: 677–680.

Carpenter, S. R., and J. F. Kitchell. 1993. *The Trophic Cascade in Lakes.* Cambridge: Cambridge University Press.

Carpenter, S. R., N. F. Caraco, D. L. Correll, R. W. Howarth, A. N. Sharpley, and V. H. Smith. 1998. Nonpoint pollution of surface waters with phosphorus and nitrogen. *Ecological Applications* 9: 559–568.

Carreiro, M. M., R. L. Sinsabaugh, D. A. Repert, and D. F. Parkhurst. 2000. Microbial enzyme shifts explain litter decay responses to simulated nitrogen deposition. *Ecology* 81: 2359–2365.

Carrillo, J. H., M. Galanter-Hastings, D. M. Sigman, and B. J. Huebert. 2002. Atmospheric deposition of inorganic and organic nitrogen and base cations in Hawaii. *Global Biogeochemical Cycles* 10.1029/2002GB001892.

Carson, H. L. 1986. Sexual selection and speciation. In *Evolutionary Processes and Theory,* eds. S. Karlin, and E. Nevo, 391–409. London: Academic Press.

Carson, H. L., and D. A. Clague. 1995. Geology and biogeography of the Hawaiian Islands. In *Hawaiian Biogeography: Evolution on a Hot Spot Archipelago,* eds. W. L. Wagner and V. A. Funk, 14–29. Washington, DC: Smithsonian Institution Press.

Cavalier, J., M. Jaramillo, D. Solis, and D. de Leon. 1997. Water balance and nutrient inputs in bulk precipitation in tropical montane cloud forest in Panama. *Journal of Hydrology* 193: 83–96.

Chadwick, O. A., and J. Chorover. 2001. The chemistry of pedogenic thresholds. *Geoderma* 100: 321–353.

Chadwick, O. A., G. H. Brimhall, and D. M. Hendricks. 1990. From a black to a gray box—a mass balance interpretation of pedogenesis. *Geomorphology* 3: 369–390.

Chadwick, O. A., C. G. Olson, D. M. Hendricks, E. F. Kelly, and R. T. Gavenda. 1994. Quantifying climatic effects on mineral weathering and neoformation in Hawaii. *Proceedings of the 15th International Soil Science Congress* 8a: 94–105.

Chadwick, O. A., L. A. Derry, P. M. Vitousek, B. J. Huebert, and L. O. Hedin. 1999. Changing sources of nutrients during four million years of ecosystem development. *Nature* 397: 491–497.

Chadwick, O. A., R. T. Gavenda, E. F. Kelly, K. Ziegler, C. G. Olson, W. C. Elliott, and D. M. Hendricks. 2003. The impact of climate on the biogeochemical functioning of volcanic soils. *Chemical Geology* 202: 193–221.

Chameides, W. L., P. S. Khasibhatla, Y. Yienger, and H. Levy II. 1994. The growth of continental-scale metro-agro-plexes, regional ozone pollution, and world food production. *Science* 264: 74–77.

Chapin, F. S. III. 1980. The mineral nutrition of wild plants. *Annual Review of Ecology and Systematics* 11: 233–260.

Chapin, F. S. III, P. M. Vitousek, and K. Van Cleve. 1986. The nature of nutrient limitation in plant communities. *American Naturalist* 127: 48–58.

Chapin, F. S. III, L. Moilanen, and K. Kielland. 1993. Preferential use of organic nitrogen for growth by a non-mycorrhizal arctic sedge. *Nature* 361: 150–153.

Chapin, F. S. III, L. R. Walker, C. L. Fastie, and L. C. Sharman. 1994. Mechanisms of primary succession following deglaciation at Glacier Bay, Alaska. *Ecological Monographs* 64: 149–175.

Chorover, J., M. J. DiChiaro, and O. A. Chadwick. 1999. Structural charge and cesium retention in a chronosequence of tephritic soils. *Soil Science Society of America Journal* 63: 169–177.

Clague, D. A. 1996. The growth and subsidence of the Hawaiian-Emperor volcanic chain. In *The Origin and Evolution of Pacific Island Biotas, New Guinea* to *Eastern Polynesia: Patterns and Processes,* eds. A. Keast, and S. E. Miller, 50. Amsterdam: SPB Academic Publishing.

Clague, D. A., and G. B. Dalrymple. 1987. The Hawaiian-Emperor volcanic chain. In *Volcanism in Hawaii,* eds. R. W. Decker, T. L. Wright, and P. M. Stauffer, 5–73. USGS Professional Paper 1350, U.S. Geological Survey, Washington, DC.

Clague, D. A., J. T. Hagstrum, D. E. Champion, and M. H. Beeson. 1999. Kilauea summit overflows: their ages and distribution in the Puna district, Hawai'i. *Bulletin Volcanologie* 61: 363–381.

Clark, D. A., S. Brown, D. W. Kicklighter, J. Q. Chambers, J. R. Thomlinson, J. Ni, and E. A. Holland. 2001. Net primary production in tropical forests: An evaluation and synthesis of existing data. *Ecological Applications* 11: 371–384.

Clark, K. L., N. M Nadkarni, D. Schaefer, and H. L. Gholz. 1998. Atmospheric deposition and net retention of ions by the canopy in a tropical montane cloud forest, Monte Verde, Costa Rica. *Journal of Tropical Ecology* 14: 27–45.

Cleveland, C. C., A. R. Townsend, D. S. Schimel, H. Fisher, R. W. Howarth, L. O. Hedin, S. S. Perakis, E. F. Latty, J. C. VonFischer, A. Elseroad, and M. F. Wasson. 1999. Global patterns of terrestrial biological nitrogen (N_2) fixation in natural ecosystems. *Global Biogeochemical Cycles* 13: 623–645.

Cochrane, O. A., and R. A. Berner. 1997. Promotion of chemical weathering by higher plants: Field observations on Hawaiian basalts. *Chemical Geology* 132: 71–85.

Coeppicus, S. 1999. Atmospheric deposition of fixed nitrogen and base cations and the volcanic source of fixed nitrogen at the Thurston lava tube, Hawai'i. MS. Thesis, University of Hawai'i, Manoa, Hawai'i.

Cole, C. V., and R. D. Heil. 1981. Phosphorus effects on terrestrial nitrogen cycling. In *Terrestrial Nitrogen Cycles: Processes, Ecosystem Strategies, and Management Impacts,* eds. F. E. Clark, and T. H. Rosswall, 363–374. *Ecological Bulletins* (Stockholm) 33.

Cole, F. R., A. C. Medeiros, L. L. Loope, and W. W. Zuehlke. 1992. Effects of the Argentine ant (*Iridomyrmex humilis*) on the arthropod fauna of high-elevation shrubland, Haleakala National Park, Maui, Hawaii. *Ecology* 73: 1313–1322.

Cooper, W. S. 1923. The recent ecological history of Glacier Bay, Alaska. *Ecology* 4: 93–128, 223–246, 355–365.

Cordell, S., and G. Goldstein. 1999. Light distribution and photosynthesis of *Metrosideros polymorpha* dominated forests at both ends of a nutrient and substrate age gradient in Hawaii. *Selbyana* 20: 350–356.

Cordell, S., G. Goldstein, D. Mueller-Dombois, D. Webb, and P. M. Vitousek. 1998. Physiological and morphological variation in *Metrosideros polymorpha,* a dominant Hawaiian tree species, along an altitudinal gradient: the role of phenotypic plasticity. *Oecologia* 113: 188–196.

Cordell, S., G. Goldstein, F. C. Meinzer, and P. M. Vitousek. 2001a. Morphological and physiological adjustment to N and P fertilization in nutrient-limited *Metrosideros polymorpha* canopy trees in Hawaii. *Tree Physiology* 21: 43–50.

Cordell, S., G. Goldstein, F. C. Meinzer, and P. M. Vitousek. 2001b. Regulation of leaf life-span and nutrient-use efficiency of *Metrosideros polymorpha* trees at two extremes of a long chronosequence in Hawaii. *Oecologia* 127: 198–206.

Cordy, R. 2000. *Exalted Sits the Chief: The Ancient History of Hawai'i Island.* Honolulu: University of Hawai'i Press.

Cowles, H. C. 1899. The ecological relations of the vegetation on the sand dunes of Lake Michigan. *Botanical Gazette* 27: 95–117.

Crews, T. E. 1993. Phosphorus regulation of nitrogen fixation in a traditional Mexican agroecosystem. *Biogeochemistry* 21: 141–166.

Crews, T. E., K. Kitayama, J. Fownes, D. Herbert, D. Mueller-Dombois, R. H. Riley, and P. M. Vitousek. 1995. Changes in soil phosphorus and ecosystem dynamics across a long soil chronosequence in Hawai'i. *Ecology* 76: 1407–1424.

Crews, T. E., H. Farrington, and P. M. Vitousek. 2000. Changes in asymbiotic, heterotrophic nitrogen fixation on leaf litter of *Metrosideros polymorpha* with long-term ecosystem development in Hawaii. *Ecosystems* 3: 386–395.

Crews, T. E., L. M. Kurina, and P. M. Vitousek. 2001. Organic matter and nitrogen accumulation and nitrogen fixation during early ecosystem development in Hawaii. *Biogeochemistry* 52: 259–279.

Crocker, R. L., and J. Major. 1955. Soil development in relation to vegetation and surface age at Glacier Bay, Alaska. *Journal of Ecology* 43: 427–448.

Cross, A. F., and W. H. Schlesinger. 1995. A literature review and evaluation of the Hedley fractionation: Applications to the biogeochemical cycle of soil phosphorus in natural ecosystems. *Geoderma* 64: 197–214.

Cuddihy, L. W., and C. P. Stone. 1990. *Alteration of Native Hawaiian Vegetation: Effects of Humans, Their Activities and Introductions.* Honolulu: University of Hawaii Cooperative National Park Resources Study Unit.

Cuevas, E. 2001. Soil versus biological controls on nutrient cycling in terra firme forests. In *The Biogeochemistry of the Amazon Basin,* eds. M. McClain, R. Victoria, and J. Richey, 53–67. New York: Oxford University Press.

Currie, W. S., J. D. Aber, W. H. McDowell, R. D. Boone, and A. H. Magill. 1996. Vertical transport of dissolved organic C and N under long-term N amendments in pine and hardwood forests. *Biogeochemistry* 35: 471–505.

Dahlgren, R. A. 1994. Soil acidification and nitrogen saturation from weathering of ammonium-bearing rock. *Nature* 368: 838–841.

D'Antonio, C. M., and T. L. Dudley. 1995. Biological invasions as agents of change on islands versus mainlands. In *Islands: Biological Diversity and Ecosystem Function,* eds. P. M. Vitousek, L. L. Loope, and H. Adsersen, 103–121. Berlin: Springer-Verlag.

D'Antonio, C. M., R. F. Hughes, and P. M. Vitousek. 2001. Factors influencing dynamics of two invasive C_4 grasses in seasonally dry Hawaiian woodlands: Resource limitation and priority effects. *Ecology* 82: 89–104.

Darwin, C. 1845. *The Voyage of the Beagle.* London: Everyman's Library, J. M. Dent.

———. 1881. The Formation of Vegetable Mould Through the Action of Worms with Observations on Their Habits. London: Murray.

Daube, B. J., K. D. Kimball, P. A. Lamar, and K. C. Weathers. 1987. Two new ground-level cloud water sampler designs which reduce rain contamination. *Atmospheric Environment* 21: 893–900.

Davidson, E. A., P. A. Matson, P. M. Vitousek, R. Riley, K. Dunkin, G. Garcia-Mendez, and J. M. Maass. 1993. Processes regulating soil emissions of NO and N_2O in a seasonally dry tropical forest. *Ecology* 74: 130–139.

Dawson, J. W., and R. L. Stemmermann. 1990. Metrosideros (Myrtaceae). In *Manual of Flowering Plants of Hawaii*, eds. W. L. Wagner, D. R. Herbst, and S. H. Sohmer, 964–970. Honolulu, Hawaii: B. P. Bishop Museum.

Denslow, J. S. 2003. Weeds in paradise: Thoughts on the invasability of tropical islands. *Annals of the Missouri Botanical Garden* 90: 119–127.

Duce, R. A., P. S. Liss, J. T. Merrill, E. L. Atlas, P. Buat-Menard, B. B. Hicks, J. M. Miller, J. M. Prospero, R. Arimoto, T. M Church, W. Ellis, J. N. Galloway, L. Hanson, T. D. Jickells, A. H. Knap, K. H. Reinhardt, B. Schneider, A. Sondine, J. J. Tokos, S. Tsunsgai, R. Wollast, and M. Zhou. 1991. The atmospheric input of trace species to the world ocean. *Global Biogeochemical Cycles* 5: 193–259.

Dymond, J., P. E. Biscaye, and R. W. Rex. 1974. Eolian origin of mica in Hawaiian soils. *Geological Society of America Bulletin* 85: 37–40.

Eggler, W. A. 1971. Quantitative studies of vegetation on sixteen lava flows on the island of Hawaii. *Tropical Ecology* 12: 66–100.

Eisele, K. A., D. S. Schimel, L. A. Kapustka, and W. J. Parton. 1989. Effects of available P and N:P ratios on non–symbiotic dinitrogen fixation in tall grass prairie soils. *Oecologia* 79: 471–474.

Elser, J. J., D. Dobberfuhl, N. A. MacKay, and J. Schampel. 1996. Organism size, life history, and N:P stoichiometry: towards a unified view of cellular and ecosystem processes. *BioScience* 46: 674–684.

Elser, J. J., W. F. Fagan, R. F. Denno, D. R. Dobberfuhl, A. Folarin, A. Huberty, S. Internlandi, S. S. Kilham, E. McCauley, K. L. Schutz, E. H. Siemann, and R. W. Sterner. 2000. Nutritional constraints in terrestrial and freshwater food webs. *Nature* 408: 578–580.

Elton, C. S. 1958. *The Ecology of Invasions by Animals and Plants*. London: Methuen.

Erskine, P. D., D. M. Bergstrom, S. Schmidt, G. R. Stewart, C. E. Tweedie, and J. D. Shaw. 1998. Subantarctic Macquarie island: a model ecosystem for studying animal-derived nitrogen sources using ^{15}N natural abundance. *Oecologia* 117: 187–193.

Evans, R. D., R. Rimer, L. Sperry, and J. Belknap. 2001. Exotic plant invasion alters nitrogen dynamics in an arid grassland. *Ecological Applications* 11: 1301–1310.

Fastie, C. L. 1995. Causes and ecosystem consequences of multiple pathways of primary succession at Glacier Bay, Alaska. *Ecology* 76: 1899–1916.

Fenn, M. E., M. A. Poth, J. D. Aber, J. S. Baron, B. T. Bormann, D. W. Johnson, D. A Lemly, S. G. McNulty, D. F. Ryan, and R. Stottlemyer. 1998. Nitrogen excess in North American ecosystems: Predisposing factors, ecosystem responses, and management strategies. *Ecological Applications* 8: 706–733.

Field, C. B., F. S. Chapin III, P. A. Matson, and H. A. Mooney. 1992. Responses of terrestrial ecosystems to the changing atmosphere: a resource-based approach. *Annual Review of Ecology and Systematics* 23: 201–235.

Fleischer, R. C., and C. E. McIntosh. 2001. Molecular systematics and biogeography of the Hawaiian avifauna. In *Evolution, Ecology, Conservation, and Management of Hawaiian Birds: A Vanishing Avifauna*, eds. J. M. Scott, S. Conant, and C. Van Riper, 51–60. *Studies in Avian Biology* 22.

Fog, K. 1988. The effect of added nitrogen on the rate of decomposition of organic matter. *Biological Review* 63: 433–462.

Forbes, C. N. 1912. Preliminary observations concerning the plant invasion in some of the lava flows of Mauna Loa, Hawaii. *Occasional Papers of the Bishop Museum* 5: 15–23.

Forbes, S. A. 1887. The lake as a microcosm. *Bulletin of the Peoria Scientific Association.*

Foster, D. R. 1988. Disturbance history, community organization and vegetation dynamics of the old-growth Pisgah Forest, southwestern New Hampshire. *Journal of Ecology* 76: 105–134.

Frost, T. M., S. R. Carpenter, A. R. Ives, and T. K. Kratz. 1995. Species compensation and complementarity in ecosystem function. In *Linking Species and Ecosystems*, eds. C. G. Jones, and J. H. Lawton, 224–239. New York: Chapman and Hall.

Galloway, J. N., and E. B. Cowling. 2002. Reactive nitrogen and the world: 200 years of change. *Ambio* 31: 64–71.

Galloway, J. N., W. H. Schlesinger, H. Levy II, A. Michaels, and J. L. Schnoor. 1995. Nitrogen fixation: atmospheric enhancement—environmental response. *Global Biogeochemical Cycles* 9: 235–252.

Gause, G. F. 1934. *The Struggle for Existence.* Baltimore: Williams and Wilkins.

Gavenda, R. T. 1992. Hawaiian quaternary paleoenvironments: A review of geological, pedological, and botanical evidence. *Pacific Science* 46: 295–307.

George, T., B. B. Bohlool, and P. W. Singleton. 1987. *Bradyrhizobium japonicum*-environment interactions, nodulation, and interstrain competition in soils along an environmental gradient. *Applied and Environmental Microbiology* 53: 1113–1117.

Giambelluca, T. W., and T. A. Schroeder. 1998. Climate. In *Atlas of Hawai'i (Third Edition)*, eds. S. P. Juvik, J. O. Juvik, and R. R. Paradise, 49–59. Honolulu: University of Hawai'i Press.

Giambelluca, T. W., M. A. Nullet, and T. A. Schroeder. 1986. *Rainfall Atlas of Hawaii.* Honolulu: State of Hawaii Department of Land and Natural Resources Report R76.

Givnish, T. J., K. J. Sytsma, J. F. Smith, and W. J. Hahn. 1995. Molecular evolution, adaptive radiation, and geographic speciation in *Cyanea* (*Campanulaceae, Lobelioideae*). In *Hawaiian Biogeography: Evolution on a Hot Spot Archipelago*, eds. W. L. Wagner and V. A. Funk, 288–337. Washington, DC: Smithsonian Institution Press.

Goodale, C. L., and J. D. Aber. 2001. The long-term effects of land-use history on nitrogen cycling in northern hardwood forests. *Ecological Applications* 11: 253–267.

Goodale, C. L., J. D. Aber, and W. H. McDowell. 2000. The long-term effects of disturbance on organic and inorganic nitrogen export in the White Mountains, New Hampshire. *Ecosystems* 3: 433–450.

Gorham, E. 1961. Factors influencing supply of major ions to inland waters, with special reference to the atmosphere. *Geological Society of America Bulletin* 72: 795–840.

Grant, P. R. 1999. *Ecology and Evolution of Darwin's Finches.* Princeton: Princeton University Press.

Graustein, W. C. 1989. ^{87}Sr/^{86}Sr ratios measure the source and flow of strontium in terrestrial ecosystems. In *Stable Isotopes in Ecological Research,* eds. P. W. Rundel, J. R. Ehleringer, and K. A. Nagy, 491–512. New York: Springer-Verlag.

Graustein, W. C., and R. L. Armstrong. 1983. The use of strontium-87/strontium-86 ratios to measure atmospheric transport into forested watersheds. *Science* 219: 289–292.

Gressel, N., J. G. McColl, C. M Preston, R. H. Newman, and R. F. Powers. 1996. Linkages between phosphorus transformations and carbon decomposition in a forest soil. *Biogeochemistry* 33: 97–123.

Gutschick, V. P. 1981. Evolved strategies in nitrogen acquisition by plants. *American Naturalist* 118: 607–637.

Hall, S. J., and P. A. Matson. 1999. Nitrogen oxide emissions after nitrogen additions in tropical forests. *Nature* 401: 152–155.

———. 2003. Nutrient status of tropical rain forests influences soil N dynamics after N additions. *Ecological Monographs* 73: 107–129

Handley, L. L., A. T. Austin, D. Robinson, C. M. Scrimgeour, J. A. Raven, T.H.E. Heaton, S. Schmidt, and G. R. Stewart. 1999. The N-15 natural abundance (delta N-15) of ecosystem samples reflects measures of water availability. *Australian Journal of Plant Physiology* 26: 185–199.

Harding, D., and J. Miller. 1982. The influence on rain chemistry of the Hawaiian volcano Kilauea. *Journal of Geophysical Research* 87: 1225–1230.

Hardy, R.W.F., R. O. Hulsten, E. K. Jackson, and R. C. Burns. 1968. The acetylene-ethylene assay for N$_2$ fixation: laboratory and field evaluation. *Plant Physiology* 43: 1185–1207.

Harrington, R. A., J. H. Fownes, F. C. Meinzer, and P. G. Scowcroft. 1995. Forest growth along a rainfall gradient in Hawaii: *Acacia koa* stand structure, productivity, foliar nutrients, and water- and nutrient-use efficiencies. *Oecologia* 102: 277–284

Harrington, R. A., J. H. Fownes, and P. M. Vitousek. 2001. Production and resource-use efficiencies in N- and P-limited tropical forest ecosystems. *Ecosystems* 4: 646–657.

Harrison, A. F., and D. R. Helliwell. 1979. A bioassay for comparing phosphorus availability in soils. *Journal of Applied Ecology* 16: 497–505.

Harte, J., and R. Shaw. 1995. Shifting dominance within a montane vegetation community: Results of a climate-warming experiment. *Science* 267: 876–880.

Hartwig, U. A. 1998. The regulation of symbiotic N$_2$ fixation: a conceptual model of N feedback from the ecosystem to the gene expression level. *Perspectives in Plant Ecology, Evolution and Systematics* 1: 92–120.

Hättenschwiler, S., A. E. Hagerman, and P. M. Vitousek. 2003. Polyphenols in litter from tropical montane forests across a wide range in soil fertility. *Biogeochemistry,* 64: 129–148.

Heath, J. A. 2001. Atmospheric Nutrient Deposition in Hawai'i: Methods, Rates, and Sources. PhD Dissertation, University of Hawai'i, Manoa, Honolulu, Hawai'i.

Heath, J. A., and B. J. Huebert. 1999. Cloudwater deposition as a source of fixed nitrogen in a Hawaiian montane forest. *Biogeochemistry* 44: 119–134.

Hedin, L. O., L. Granat, G. E. Likens, T. A. Buishand, J. N. Galloway, T. J. Butlers, and H. Rodhe. 1994. Steep declines in atmospheric base cations in regions of Europe and North America. *Nature* 367: 351–354.

Hedin. L. O., J. J. Armesto, and A. H. Johnson. 1995. Patterns of nutrient loss from unpolluted, old-growth temperate forests: evaluation of biogeochemical theory. *Ecology* 76: 493–509.

Hedin, L. O., J. C. von Fischer, N. E. Ostrom, B. P. Kennedy, M. G. Brown, and G. P. Robertson. 1998. Thermodynamic constraints on nitrogen transformations and other biogeochemical processes at soil-stream interfaces. *Ecology* 79: 684–703.

Hedin, L. O., P. M. Vitousek, and P. A. Matson. 2003. Pathways and implications of nutrient losses during four million years of tropical forest ecosystem development. *Ecology* 84: 2231–2255.

Hedley, M. J., J.W.B. Stewart, and B. S. Chauhan. 1982. Changes in inorganic and organic soil phosphorus fractions induced by cultivation practices and laboratory incubations. *Soil Science Society of America Journal* 46: 970–976.

Heneghan, L., D. C. Coleman, X. Zou, D. A. Crossley Jr, and B. L. Haines. 1999. Soil microarthropod contributions to decomposition dynamics: tropical-temperate comparisons of a single substrate. *Ecology* 80: 1873–1882.

Herbert, D. A., and J. H. Fownes. 1995. Phosphorus limitation of forest leaf area and net primary productivity on a weathered tropical soil. *Biogeochemistry* 29: 223–235.

———. 1999. Forest productivity and efficiency of resource use across a chronosequence of tropical montane soils. *Ecosystems* 2: 242–254.

Herbert, D. A., J. H. Fownes, and P. M. Vitousek. 1999. Hurricane damage and recovery of a Hawaiian forest: effects of increased nutrient availability on ecosystem resistance and resilience. *Ecology* 80: 908–920.

Hicks, B. B., D. D. Baldocci, J.R.P. Hosker, B. A. Hutchinson, D. R. Matt, R. T. McMillen, and L. C. Satterfield. 1985. On the use of monitored air concentrations to infer dry deposition. NOAA Technical Memorandum ERL-ARL-141.

Hiremath, A. J, and J. J. Ewel. 2001. Ecosystem nutrient use efficiency, productivity, and nutrient accrual in model tropical communities. *Ecosystems* 4: 669–683.

Hobbie, S. E. 1992. Effects of plant species on nutrient cycling. *Trends in Ecology and Evolution* 7: 336–339.

———. 1995. Direct and indirect effects of plant species on biogeochemical processes in arctic ecosystems. In *Arctic and Alpine Biodiversity: Patterns, Causes and Ecosystem Consequences,* eds. F. S. Chapin II, and C. Körner, 213–214. Berlin: Springer-Verlag.

———. 2000. Interactions between litter lignin and soil nitrogen availability during leaf litter decomposition in a Hawaiian montane rainforest. *Ecosystems* 3: 484–494.

Hobbie, S. E., and P. M. Vitousek. 2000. Nutrient limitation of decomposition in Hawaiian forests. *Ecology* 81: 1867–1877.

Högberg, P., and C. Johannison. 1993. [15]N abundance of forests is correlated with losses of nitrogen. *Plant and Soil* 157: 147–150.

Holcomb, R. T., P. W. Reiners, B. K. Nelson, and N.-L.E. Sawyer. 1997. Evidence for two shield volcanoes exposed on the island of Kaua'i, Hawai'i. *Geology* 25: 811–814.

Holland, E. A., W. J. Parton, J. K.Detling, and D. L. Coppock. 1992. Physiological responses of plant populations to herbivory and their consequences for ecosystem nutrient flow. *American Naturalist* 140: 685–706.

Holland, E. A., F. J. Dentener, B. H. Braswell, and J. M. Sulzman. 1999. Contemporary and pre-industrial global reactive nitrogen budgets. *Biogeochemistry* 46: 1–37.

Holloway, J. M., R. A. Dahlgren, B. Hansen, and W. H. Casey. 1998. Contribution of bedrock nitrogen to high nitrate concentrations in stream water. *Nature* 395: 785–788.

Hooper, D. U., and P. M. Vitousek. 1998. Effects of plant composition and diversity on nutrient cycling in serpentine grassland. *Ecological Monographs* 68: 121–149.

Hooper, D. U., J. J. Ewel, J. P. Grime, A. Hector, P. Inchausti, S. Lavorel, J. Lawton, D. Lodge, M. Loreau, S. Naeem, B. Schmid, H. Setälä, A. J. Symstad, J. Vandermeer, and D. A. Wardle. Effects of biodiversity on ecosystem functioning: A consensus of current knowledge and needs for future research. *Ecological Applications,* in press.

Hostetler, S. W. and P. U. Clark. 2000. Tropical climate at the last glacial maximum inferred from glacier mass-balance modeling. *Science* 290: 1747–1750.

Hotchkiss, S. C. 1998. Quaternary vegetation and climate of Hawai'i. PhD Dissertation, University of Minnesota, St Paul.

———. Quaternary history from the U.S. tropics. In *The Quarternary Period in the United States,* eds. A. Gillespie, S. Poter, and B. Atwater. In Press. Amsterdam: Elsevier Publishers.

Hotchkiss, S., and J. O. Juvik. 1999. A late-quaternary pollen record from Ka'au Crater, O'ahu, Hawai'i. *Quaternary Research* 52: 115–128.

Hotchkiss, S. C., P. M. Vitousek, O. A. Chadwick, and J. P. Price. 2000. Climate cycles, geomorphological change, and the interpretation of soil and ecosystem development. *Ecosystems* 3: 522–533.

Howarth, R. W., and J. J. Cole. 1985. Molybdenum availability, nitrogen limitation, and phytoplankton growth in natural waters. *Science* 229: 653–655.

Howarth, R. W., F. Chan, and R. Marino. 1999. Do top-down and bottom-up controls interact to exclude nitrogen-fixing cyanobacteria from the plankton of estuaries? An exploration with a simulation model. *Biogeochemistry* 46: 203–231.

Huebert, B., P. Vitousek, J. Sutton, T. Elias, J. Heath, S. Coeppicus, S. Howell, and B. Blomquist. 1999. Volcano fixes nitrogen into plant-available forms. *Biogeochemistry* 47: 111–118.

Hughes, R. F., P. M. Vitousek, and J. T. Tunison. 1991. Effects of invasion by fire-enhancing C_4 grasses on native shrubs in Hawaii Volcanoes National Park. *Ecology* 72: 743–747.

Hunt, H. W., J. W. B. Stewart, and C. V. Cole. 1983. A conceptual model for the interactions of carbon, nitrogen, phosphorus, and sulfur in grasslands. In *The Major Biogeochemical Cycles and Their Interactions,* eds. B. Bolin, and R. B. Cook, 303–326. New York: John Wiley and Sons.

Hunt, H. W., E. R. Ingham, D. C. Coleman, E. T. Elliott, and C.P.P. Reid. 1988. Nitrogen limitation of production and decomposition in prairies, mountain meadow, and pine forest. *Ecology* 69: 1009–1016.

Iiyama, K., and A.F.A. Wallis. 1990. Determination of lignin in herbaceous plants by an improved acetyl bromide procedure. *Journal of the Science of Food and Agriculture* 51: 145–161.

Jackson, M. L., T. W. Levelt, J. K. Syers, R. W. Rex, R. N. Clayton, G. D. Sherman, and G. Uehara. 1971. Geomorphological relationships of tropospherically-derived quartz in the soils of the Hawaiian Islands. *Soil Science Society of America Proceedings* 35: 515–525.

James, E. K. 2000. Nitrogen fixation in epiphytic and associative symbiosis. *Field Crops Research* 65: 197–209.

James, H. F. 1995. Prehistoric extinctions and ecological changes on oceanic islands. In *Islands: Biological Diversity and Ecosystem Function*, eds. P. M. Vitousek, L. L. Loope, and H. Adsersen, 87–102. Berlin: Springer-Verlag.

Jenkinson, D. S., N. J. Bradbury, and K. Coleman. 1994. How the Rothamsted classical experiments have been used to develop and test models for the turnover of carbon and soil nitrogen. In *Long-term Experiments in Agricultural and Ecological Studies*, eds. R. A. Leigh, and A. E. Johnson, 117–138. Oxford: CAB International.

Jenny, H. 1941. *Factors of Soil Formation: A System of Quantitative Pedology*. New York: McGraw-Hill.

———. 1980. *Soil Genesis with Ecological Perspectives*. New York: Springer-Verlag.

Jenny, H., S. P. Gessell, and F. T. Bingham. 1949. Comparative study of decomposition rates of organic matter in temperate and tropical regions. *Soil Science* 68: 419–432.

Jobbagy, E. G., and R. B. Jackson. 2001. The distribution of soil nutrients with depth: Global patterns and the imprint of plants. *Biogeochemistry* 53: 51–77.

Johnson, C. M., D. J. Zarin, and A. H. Johnson. 2000. Post-disturbance aboveground biomass accumulation in global secondary forests. *Ecology* 81: 1395–1401.

Juvik, J. O. 1998. Biogeography. In *Atlas of Hawai'i (Third Edition)*, eds. S. P. Juvik, J. O. Juvik, and T. R. Paradise, 103–107. Honolulu: University of Hawai'i Press.

Juvik, J. O., and D. Nullet. 1993. Relationships between rainfall, cloud-water interception, and canopy throughfall in a Hawaiian montane forest. In *Tropical Montane Cloud Forests*, eds. L. S. Hamilton, J. O. Juvik, and F. N. Scatena, 102–114. Honolulu: East-West Center.

———. 1994. A climate transect through tropical montane rainforest in Hawaii. *Journal of Applied Meteorology* 33: 1304–1312.

Juvik, S. P., J. O. Juvik, and T. R. Paradise (eds.). 1998. *Atlas of Hawai'i (Third Edition)*. Manoa, Honolulu: University of Hawai'i Press.

Kaneshiro, K. Y. 1995. Evolution, speciation, and the genetic structure of island populations. In *Islands: Biological Diversity and Ecosystem Function*, eds. P. M. Vitousek, L. L. Loope, and H. Adsersen, 23–33. Berlin: Springer-Verlag.

Karl, D. M. 1999. A sea of change: Biogeochemical variability in the North Pacific subtropical gyre. *Ecosystems* 2: 181–214.

Karl, D. M., A. Michaels, B. Bergman, D. Capone, E. Carpenter, R. Letelier, F. Lipschultz, H. Paerl, D. Sigman, and L. Stal. 2002. Dinitrogen fixation in the world's oceans. *Biogeochemistry* 57: 47–98.

Kelly, E. F., O. A. Chadwick, and T. Hilinski. 1998. The effect of plants on mineral weathering. *Biogeochemistry* 42: 21–53

Kennedy, M. J., O. A. Chadwick, P. M. Vitousek, L. A. Derry, and D. M. Hendricks. 1998. Replacement of weathering with atmospheric sources of base cations during ecosystem development, Hawaiian Islands. *Geology* 26: 1015–1018.

Kerr, B., M. A. Riley, M. W. Feldman, and B. H. Bohannan. 2002. Local dispersal promotes biodiversity in a real-life game of rock-paper-scissors. *Nature* 418: 171–174.

Killingbeck, K. T. 1996. Nutrients in senesced leaves: keys to the search for potential resorption and resorption proficiency. *Ecology* 77: 1716–1727.

Kirch, P. V. 1985. *Feathered Gods and Fishhooks: An Introduction to Hawaiian Archaeology and Prehistory*. Honolulu: University of Hawaii Press.

———. 1994. *The Wet and the Dry: Irrigation and Agricultural Intensification in Polynesia*. Chicago: University of Chicago Press.

———. 1997. Microcosmic histories: island perspectives on "global" change. *American Anthropologist* 99: 30–42.

———. 2000. *On the Road of the Winds: An Archaeological History of the Pacific Islands Before European Contact*. Berkeley: University of California Press.

Kirch, P. V., and R. C. Green. 2001. *Havaiki, Ancestral Polynesia: An Essay in Historical Anthropology*. Cambridge: Cambridge University Press.

Kitayama, K. and S.-I. Aiba. 2002. Ecosystem structure and productivity of tropical rain forests along altitudinal gradients with contrasting soil phosphorus pools on Mount Kinabalu, Borneo. *Journal of Ecology* 90: 37–51.

Kitayama, K., and D. Mueller-Dombois. 1994. An altitudinal transect analysis of the windward vegetation on Haleakala, a Hawaiian island mountain: (1) climate and soils. *Phytocoenologia* 24: 41–133.

———. 1995. Vegetation changes along gradients of long-term soil development in the Hawaiian montane rainforest zone. *Vegetatio* 120: 1–20.

Kitayama, K., E.A.G. Schuur, D. R. Drake, and D. Mueller-Dombois. 1997. Fate of a wet montane forest during soil aging in Hawaii. *Journal of Ecology* 85: 669–679.

Kitayama, K., N. Majalap-Lee, and S. Aiba. 2000. Soil phosphorus fractionation and phosphorus-use efficiencies of tropical rain forests along altitudinal gradients of Mount Kinabalu, Borneo. *Oecologia* 123: 342–349.

Krebs, H. A. 1975. The August Krogh principle: "for many problems there is an animal on which it can most conveniently be studied." *Journal of Experimental Zoology* 194: 221–226.

Kroeze, C., A. Mosier, and L. Bouwman. 1999. Closing the N_2O budget: A retrospective analysis. *Global Biogeochemical Cycles* 13: 1–8.

Krogh, A. 1929. Progress of physiology. *American Journal of Physiology* 90: 243–251.

Kurina, L. M., and P. M. Vitousek. 1999. Controls over the accumulation and decline of a nitrogen fixing lichen, *Stereocaulon vulcani,* on young Hawaiian lava flows. *Journal of Ecology* 87: 784–799.

———. 2001. Nitrogen fixation rates of *Stereocaulon vulcani* on young Hawaiian lava flows. *Biogeochemistry* 55: 179–194.

Kurtz, A. C., L. A. Derry, O. A. Chadwick, and M. J. Alfano. 2000. Refractory element mobility in volcanic soils. *Geology* 28: 683–686.

Kurtz, A. C., L. A. Derry, and O. A. Chadwick. 2001. Accretion of Asian dust to Hawaiian soils: isotopic, elemental, and mineral mass balances. *Geochimica et Cosmochimica Acta* 65: 1971–1983.

Lajtha, K., and W. H. Schlesinger. 1988. The biogeochemistry of phosphorus cycling and phosphorus availability along a desert soil chronosequence. *Ecology* 69: 24–39.

Lajtha, K., B. Seely, and I. Valiela. 1995. Retention and leaching losses of atmospherically derived nitrogen in the aggrading coastal watershed of Waquoit Bay, Massachusetts. *Biogeochemistry* 28: 33–54.

Lavelle, P. 1997. Faunal activities and soil processes: Adaptive strategies that determine ecosystem function. *Advances in Ecological Research* 27: 93–132.

Le Maitre, D. C., B. W. Van Wilgen, R. A. Chapman, and D. McKelly. 1995. Invasive plants and water resources in the western Cape province, South Africa: Modeling the consequences of a lack of management. *Journal of Applied Ecology* 33: 161–172.

Lee, J. A., R. Harmer, and R. Ingaciuk. 1983. Nitrogen as a limiting factor in plant communities. In *Nitrogen as an Ecological Factor*, eds. J. A. Lee, S. McNeill, and I. H. Rorison, 95–112. Oxford: Blackwell Scientific.

Leinen, M., J. M. Prospero, E. Arnold, and M. Blank. 1994. Mineralogy of aeolian dust reaching the North Pacific Ocean. I. Sampling and analysis. *Journal of Geophysical Research* 99: 21017–21024.

Lerdau, M. T., J. W. Munger, and D. J. Jacob. 2000. The NO_2 flux conundrum. *Science* 289: 2291–2292.

Lichter, J. 1998. Rates of weathering and chemical depletion in soils across a chronosequence of Lake Michigan sand dunes. *Geoderma* 85: 255–282.

Likens, G. E., F. H. Bormann, N. M Johnson, D. W. Fisher, and R. S. Pierce. 1970. Effects of forest cutting and herbicide treatment on nutrient budgets in the Hubbard Brook watershed-ecosystem. *Ecological Monographs* 40: 23–47.

Likens, G. E., F. H. Bormann, R. S. Pierce, J. S. Eaton, and N. M. Johnson. 1977. *Biogeochemistry of a Forested Ecosystem*. New York: Springer-Verlag.

Lindemann, R. L. 1942. The trophic-dynamic aspect of ecology. *Ecology* 23: 399–417.

Llinás, R. R. 1999. *The Squid Giant Synapse*. New York: Oxford University Press.

Lloyd, J., M. I. Bird, E. M. Veenendaal, and B. Kruijt. 2001. Should phosphorus availability be constraining moist tropical forest responses to increasing CO_2 concentrations? In *Global Biogeochemical Cycles in the Climate System*, eds. E. D. Schulze, S. P. Harrison, M. Heimann, E. A. Holland, J. Lloyd, I. C. Prentice, and D. Schimel, 96–114. San Diego: Academic Press.

Lockwood, J. P. 1995. Mauna Loa eruptive history: the preliminary radiocarbon record. *Geophysical Monographs* 92: 81–94.

Lohse, K. 2002. Hydrological and Biogeochemical Controls on Nitrogen Losses from Tropical Forests: Responses to Anthropogenic Nitrogen Additions. Ph.D. dissertation, University of California, Berkeley.

Loope, L. L. 1998. Hawaii and the Pacific islands. In *Status and Trends in the Nation's Biological Resources,* eds. M. J. Mac, C. E. Opler, C.E.P. Haecker, and P. D. Doran, 747–774. Reston, Virginia: U.S. Department of the Interior, U.S. Geological Survey.

Loope, L. L., and T. W. Giambelluca. 1998. Vulnerability of island tropical cloud forests to climate change, with special reference to East Maui, Hawaii. *Climatic Change* 39: 503–517.

Ludwig, K. R., B. J. Szabo, J. G. Moore, and K. R. Simmons. 1991. Crustal subsidence rate off Hawai'i, determined from $^{234}U/^{238}U$ ages of drowned coral reefs. *Geology* 19: 171–174.

MacCaughey, V. 1917. Vegetation of Hawaiian lava flows. *Botanical Gazette* 64: 386–420.

MacDonald, G. A., A. T. Abbot, and F. L. Peterson. 1983. *Volcanoes in the Sea: The Geology of Hawaii.* Honolulu: University of Hawaii Press.

Mack, M. C., C. M. D'Antonio, and R. E. Ley. 2001. Alteration of ecosystem N dynamics by exotic plants: a case study of C_4 grasses in Hawaii. *Ecological Applications* 11: 1323–1335.

Martinelli, L. A., M. C. Piccolo, A. R. Townsend, P. M. Vitousek, E. Cuevas, W. H. McDowell, G. P. Robertson, O. C. Santos, and K. Treseder. 1999. Nitrogen stable isotope composition of leaves and soil: tropical versus temperate forests. *Biogeochemistry* 46: 45–65.

Marrs, R. H., J. Proctor, A. Heaney, and M. D. Mountford. 1988. Changes in soil nitrogen-mineralization and nitrification along an altitudinal transect in tropical rainforest in Costa Rica. *Journal of Ecology* 76: 466–482.

Matson, P. A. 1990. Plant-soil interactions in primary succession at Hawaii Volcanoes National Park. *Oecologia* 85: 241–246.

Matson, P. A., and R. D. Boone. 1984. Natural disturbance and nitrogen mineralization: Wave-form dieback of mountain hemlock in the Oregon Cascades. *Ecology* 65: 1511–1516.

Matson, P. A., and A. Goldstein. 2000. Biogenic trace gas exchanges. In *Methods in Ecosystem Science,* eds. O. E. Sala, R. B. Jackson, H. A. Mooney, and R. W. Howarth, 235–248. New York: Springer-Verlag.

Matson, P. A., and P. M. Vitousek. 1987. Cross-system comparison of soil nitrogen transformations and nitrous oxide fluxes in tropical forests. *Global Biogeochemical Cycles* 1: 163–170.

Matson, P. A., C. Billow, S. Hall, and J. Zachariassen. 1996. Fertilization practices and soil variations control nitrogen oxide emissions from tropical sugar cane. *Journal of Geophysical Research* 101: 18533–18545.

Matzek, V. A., and P. M. Vitousek. 2003. Nitrogen fixation in bryophytes, lichens, and decaying wood along a soil age gradient in Hawaiian montane rainforest. *Biotropica* 35: 12–19.

McDowell, W. H., and C. E. Asbury. 1994. Export of carbon, nitrogen, and major ions from three tropical montane watersheds. *Limnology and Oceanography* 39: 111–125.

McGill, W. B., and C. V. Cole. 1981. Comparative aspects of cycling of organic C, N, S, and P through soil organic matter. *Geoderma* 26: 267–286.

McKey, D. 1994. Legumes and nitrogen: the evolutionary ecology of a nitrogen-demanding lifestyle. In *Advances in Legume Systematics: Part 5—The Nitrogen Factor,* eds. J. I. Sprent, and D. McKey, 211–228. Kew, England: Royal Botanic Gardens.

Medina, E., and E. Cuevas. 1994. Mineral nutrition: Humid tropical forests. *Progress in Botany* 55: 115–127.

Melillo, J. M., P. A. Steudler, J. D. Aber, K. Newkirk, H. Lux, F. P. Bowles, C. Catricala, A. Magill, T. Ahrens, and S. Morrisseau. 2002. Soil warming and carbon-cycle feedbacks to the climate system. *Science* 298: 2173–2176.

Miller, E. K., J. D. Blum, and A. J. Friedland. 1993. Determination of soil exchangeable-cation loss and weathering rates using Sr isotopes. *Nature* 362: 438–441.

Miller, H. G. 1981. Forest fertilization: Some guiding concepts. *Forestry* 54: 157–167.

Moore, J. G., and D. A. Clague. 1992. Volcano growth and evolution of the Island of Hawaii. *Geological Society of America Bulletin* 104: 1471–1484.

Moore, J. G., W. R. Normark, and R. T. Holcomb. 1994. Giant Hawaiian underwater landslides. *Science* 264: 46–47.

Mueller-Dombois, D. 1986. Perspectives for an etiology of stand-level dieback. *Annual Review of Ecology and Systematics* 17: 221–243.

———. 1992. Distributional dynamics in the Hawaiian vegetation. *Pacific Science* 46: 221–231.

Mueller-Dombois, D., and F. R. Fosberg. 1998. *Vegetation of the Tropical Pacific Islands.* New York: Springer-Verlag.

Mueller-Dombois, D., K. W. Bridges, and H. L. Carson (eds.). 1981. *Island Ecosystems: Biological Organization in Selected Hawaiian Communities.* Stroudsburg, PA: Hutchinson Ross.

Mueller-Dombois, D., P. M. Vitousek, and K. W. Bridges. 1984. Canopy dieback and ecosystem processes in Pacific forests. *Hawaii Botanical Science Paper 44,* 100 p. University of Hawaii.

Nakai, S., A. N. Halliday, and D. K. Rea. 1993. Provenance of dust in the Pacific Ocean. *Earth and Planetary Science Letters* 119: 143–157.

Näsholm, T., A. Ekblad, A. Nordin, R. Giesler, M. Högberg, and P. Högberg. 1998. Boreal forest plants take up organic nitrogen. *Nature* 392: 914–916.

National Research Council. 1999. *Our Common Journey: A Transition toward Sustainability.* Washington, D.C.: National Academy of Sciences Press.

Neff, J. C., S. E. Hobbie, and P. M. Vitousek. 2000. Nutrient and mineralogical control on dissolved organic C, N and P fluxes and stoichiometry in Hawaiian soils. *Biogeochemistry* 51: 283–302.

Newman, D. K., and J. F. Banfield. 2002. Geomicrobiology: How molecular-scale interactions underpin biogeochemical systems. *Science* 296: 1071–1077.

Newman, E. I. 1995. Phosphorus inputs to terrestrial ecosystems. *Journal of Ecology* 83: 713–726.

Northrup, R. R., R. A. Dahlgren, and J. G. McColl. 1998. Polyphenols as regulators of plant-litter-soil interactions: a positive feedback. *Biogeochemistry* 42: 189–220.

Nusslein, K., and J. M. Tiedje. 1998. Characterization of the dominant and rare members of a young Hawaiian soil bacterial community with small-subunit ribosomal DNA amplified from DNA—fractionated on the basis of its guanine and cytosine composition. *Applied and Environmental Microbiology* 64: 1283–1289.

Nye, P. H. and P. B. Tinker. 1977. *Solute Movement in the Soil-Root System.* Berkeley: University of California Press.

Olander, L. P. 2002. Geochemical and biological control over short-term phosphorus dynamics in tropical soils. Ph.D. Dissertation, Stanford University, Stanford, CA.

Olander, L. P., and P. M. Vitousek. 2000. Regulation of soil phosphatase and chitinase activity by N and P availability. *Biogeochemistry* 49: 175–190.

———. Biological and geochemical sinks for phosphorus in a wet tropical forest soil. *Ecosystems,* in press.

Oren, R., D. S. Ellsworth, K. H. Johnson, N. Phillips, B. E. Ewers, C. Maier, K.V.R. Schafer, H. McCarthy, G. Hendrey, S. G. McNulty, and G. G. Katul. 2001. Soil fertility limits carbon sequestration by forest ecosystems in a CO_2-enriched world. *Nature* 411: 469–472.

Ostertag, R. 2001. The effects of nitrogen and phosphorus availability on fine root dynamics in Hawaiian montane forests. *Ecology* 82: 485–499.

Ostertag, R., and S. E. Hobbie. 1999. Early stages of root and leaf decomposition in Hawaiian forests: Effects of nutrient availability. *Oecologia* 121: 564–573.

Ostertag, R., and J. Verville. 2002. Fertilization with nitrogen and phosphorus increases abundance of non-native species in Hawaiian montane forests. *Plant Ecology* 162: 77–90.

Parrington, J. R., W. H. Zoller, and N. K. Aras. 1983. Asian dust: Seasonal transport to the Hawaiian Islands. *Science* 220: 195–197.

Parton, W. J., J.M.O Scurlock, D. S. Ojima, T. G. Gilmanov, R. J. Scholes, D. S. Schimel, T. Kirchner, J.-C. Menaut, T. Seastedt, E. Garcia-Montoya, A. Kamnalrut, and J. I. Kinyamario. 1993. Observations and modeling of biomass and soil organic matter dynamics for the grassland biome world-wide. *Global Biogeochemical Cycles* 7: 785–809.

Pastor, J. P., and W. M. Post. 1986. Influence of climate, soil moisture, and succession on forest carbon and nitrogen cycles. *Biogeochemistry* 2: 3–27.

Paul, E. A., and F. E. Clark. 1996. *Soil Microbiology and Biochemistry.* San Diego: Academic Press.

Pearson, H. L. 1998. Regulation of symbiotic nitrogen fixation in a tropical leguminous tree. PhD Dissertation, Stanford University, Stanford, CA.

Pearson, H. L., and P. M. Vitousek. 2001. Symbiotic nitrogen fixation in regenerating stands of *Acacia koa,* a tropical leguminous tree. *Ecological Applications* 11: 1381–1394.

———. 2002. Nitrogen and phosphorus dynamics and symbiotic nitrogen fixation across a substrate age gradient in Hawai'i. *Ecosystems* 5: 587–596.

Perakis, S. S., and L. O. Hedin. 2001. Fluxes and fate of inorganic nitrogen in an unpolluted old-growth rainforest in southern Chile. *Ecology* 82: 2245–2260.

———. 2002. Nitrogen loss from unpolluted South American forests mainly via dissolved organic compounds. *Nature* 415: 416–419.

Peterjohn, W. T., J. M. Melillo, P. A. Steudler, K. M. Newkirk, F. P. Bowles, and J. D. Aber. 1994. Responses of trace gas fluxes and N availability to experimentally elevated soil temperatures. *Ecological Applications* 4: 617–625.

Petit, J. R., J. Jouzel, D. Raynaud, N. I. Barkov, J. M. Barnola, I. Basile, M. Bender, J. Chappellza, M. Davis, G. Delaygue, M. Delmotte, V. M. Kotlyakov, M. Legrand, V. Y. Lipenkov, C. Lorius, L. Pepin, C. Ritz, E. Saltzman, and M. Stievenard. 1999. Climate and atmospheric history of the past 420,000 years from the Vostok ice core, Antarctica. *Nature* 399: 429–436.

Petrov, D. A., T. A. Sangster, J. S. Johnston, D. L. Hartl, and K. L. Shaw. 2000. Evidence for DNA loss as a determinant of genome size. *Science* 287: 1060–1062.

Pickett, S.T.A. 1989. Space-for-time substitution as an alternative to long-term studies. In *Long-term Studies in Ecology,* ed. G. E. Likens, 110–135. New York: Springer-Verlag.

Porder, S., A. Paytan, and P. M. Vitousek. Erosion and landscape development affect plant nutrient status in the Hawaiian Islands. Submitted.

Porter, S. C. 1979. Hawaiian glacial ages. *Quaternary Research* 12: 161–187.

Powers, R. F. 1980. Mineralizable soil nitrogen as an index of nitrogen availability to forest trees. *Soil Science Society of America Journal* 44: 1314–1320.

Prescott, C. E. 1995. Does nitrogen availability control rates of litter decomposition in forests? *Developments in Plant and Soil Sciences* 62: 83–88.

Price, J. P. 2002. Paleogeography and Floristic Biogeography of the Hawaiian Islands. Ph.D. Dissertation, University of California, Davis. Davis, California.

Price, J. P., and D. Clague. 2002. How old is the Hawaiian biota? Geology and phylogeny suggest recent divergence. *Proceedings of the Royal Society of London, Series B* Biological Science 269: 2429–2435.

Qualls, R. G., B. L. Haines, and W. T. Swank. 1991. Fluxes of dissolved organic nutrients and humic substances in a deciduous forest. *Ecology* 72: 254–266.

Qualls, R. G., B. L. Haines, W. T. Swank, and S. W. Tyler. 2000. Soluble organic and inorganic fluxes in clearcut and mature deciduous forests. *Soil Science Society of America Journal* 64: 1068–1077.

Raich, J. W., and K. J. Nadelhoffer. 1989. Belowground carbon allocation in forest ecosystems: global trends. *Ecology* 70: 1346–1354.

Raich, J. W., A. E. Russell, T. E. Crews, H. Farrington, and P. M. Vitousek. 1996. Both nitrogen and phosphorus limit plant production on young Hawaiian lava flows. *Biogeochemistry* 32: 1–14.

Raich, J. W., A. E. Russell, and P. M. Vitousek. 1997. Primary production and ecosystem development along an elevational gradient in Hawaii. *Ecology* 78: 707–721.

Raich, J. W., W. J. Parton. A. E. Russell, R. L. Sanford Jr., and P. M. Vitousek. 2000. Analysis of factors regulating ecosystem development on Mauna Loa using the Century model. *Biogeochemistry* 51: 161–191.

Rastetter, E. B., and G. R. Shaver. 1992. A model of multiple element limitation for acclimating vegetation. *Ecology* 73: 1157–1174.

Rastetter, E. B., G. I. Ågren, and G. R. Shaver. 1997. Responses of N-limited ecosystems to increased CO_2: a balanced-nutrition, coupled-element-cycles model. *Ecological Applications* 7: 444–460.

Rastetter, E. B., P. M. Vitousek, C. Field, G. R. Shaver, D. Herbert, and G. I. Ågren. 2001. Resource optimization and symbiotic N fixation. *Ecosystems* 4: 369–388.

Rea, D. K. 1994. The paleoclimate record provided by eolian deposition in the deep sea: The geological history of the wind. *Reviews of Geophysics* 2: 159–195.

Redfield, A. C. 1958. The biological control of chemical factors in the environment. *American Scientist* 46: 205–221.

Reich, P. B., M. B. Walters, and D. S. Ellsworth. 1992. Leaf life-span in relation to leaf, plant, and stand characteristics among diverse ecosystems. *Ecological Monographs* 62: 365–392.

———. 1997. From tropics to tundra: Global convergence in plant functioning. *Proceedings of the National Academy of Sciences U.S.A.* 94: 13730–13734.

Reiners, W. A. 1981. Nitrogen cycling in relation to ecosystem succession. In *Terrestrial Nitrogen Cycles: Processes, Ecosystem Strategies, and Management Impacts,* eds. F. E. Clark, and T. H. Rosswall, 507–528. *Ecological Bulletins* (Stockholm) 33.

———. Complementary models for ecosystems. *American Naturalist* 127: 59–73.

Rhoades, D. F. 1979. Evolution of plant chemical defenses against herbivores. In *Herbivores: Their Interaction with Secondary Plant Metabolites,* eds. G. Rosenthal, and D. H. Janzen, 4–54. New York: Academic Press.

Richter, D. D., and D. Markewitz. 2001. *Understanding Soil Change: Soil Sustainability over Millenia, Centuries, and Decades.* Cambridge: Cambridge University Press.

Riley, R. H. 1996. Process-level regulation of nitrogen trace gas flux in Hawaiian montane rainforest. *Soil Biology and Biochemistry* 28: 1251–1260.

Riley, R. H. and P. M. Vitousek. 1995. Nutrient dynamics and trace gas flux during ecosystem development in Hawaiian montane rainforest. *Ecology* 76: 292–304.

———. 2000. Hurricane effects on nitrogen trace gas emissions in Hawaiian montane rainforest. *Biotropica* 32: 751–756.

Ritchie, M. E., D. Tilman, and J.M.H. Knops. 1998. Herbivore effects on plant and nitrogen dynamics in oak savanna. *Ecology* 79: 165–177.

Robertson, G. P. 1989. Nitrification and denitrification in humid tropical ecosystems: potential controls on nitrogen retention. In *Mineral Nutrients in Tropical Forest and Savanna Ecosystems,* ed. J. Proctor, 55–69. Oxford, United Kingdom: Blackwell Scientific.

Robichaux, R. H., G. D. Carr, M. Liebman, and R. Pearcy. 1990. Adaptive radiation of the Hawaiian silversword alliance (*Compositae-Madiinae*): Ecological, morphological, and physiological diversity. *Annals of the Missouri Botanical Garden* 77: 64–72.

Rothstein, D. E., P. M. Vitousek, and B. L Simmons. An exotic tree accelerates decomposition and nutrient turnover in a Hawaiian montane rainforest. *Ecosystems,* in press.

Roughgarden, J. 1995. Vertebrate patterns on islands. In *Islands: Biological Diversity and Ecosystem Function,* eds. P. M. Vitousek, L. L. Loope, and H. Adsersen, 51–56. Berlin: Springer-Verlag.

Russell, A. E., J. W. Raich, and P. M. Vitousek. 1998. The ecology of the climbing fern, *Dicranopteris linearis*, on windward Mauna Loa, Hawai'i. *Journal of Ecology* 86: 765–779.

Rustad, L. E., J. L. Campbell, G. M. Marion, R. J. Norby, M. J. Mitchell, A. E. Hartley, J.H.C. Cornelissen, and J. Gurevitch. 2001. A meta-analysis of the response of soil respiration, net nitrogen mineralization, and aboveground plant growth to experimental ecosystem warming. *Oecologia* 126: 543–562.

Ryan, M. G., J. M. Melillo, and A. Ricca. 1989. A comparison of methods for determining proximate carbon fractions of forest litter. *Canadian Journal of Forest Research* 20: 166–171.

Sanchez, P. A. 1976. *Properties and Management of Soils in the Tropics*. New York: John Wiley and Sons.

Sansone, F. J., C. R. Benitez-Nelson, J. A. Resing, E. H. DeCarlo, S. M. Vink, J. H. Carrillo, and B. J. Huebert. 2001. Geochemistry of atmospheric aerosols generated from lava-seawater interactions. *Geophysical Research Letters* 29. 10.1029/2001GLO13882.

Scatena, F. N., and A. E. Lugo. 1995. Geomorphology, disturbance, and the soil and vegetation of two subtropical wet steepland watersheds of Puerto Rico. *Geomorphology* 13: 199–213.

Schimel, D. S., M. A. Stillwell, and R. G. Woodmansee. 1985. Biogeochemistry of C, N, and P in a soil catena of the shortgrass steppe. *Ecology* 66: 276–282.

Schimel, D. S., B. H. Brassell, and W. J. Parton. 1997. Equilibration of the terrestrial water, nitrogen, and carbon cycles. *Proceedings of the National Academy of Sciences* 94: 8280–8283.

Schimel, J. P. and M. N. Weintraub. 2003. The implications of exoenzyme activity on microbial carbon and nitrogen limitation in soil: a theoretical model. *Soil Biology and Biochemistry* 35: 549–563.

Schindler, D. W. 1977. Evolution of phosphorus limitation in lakes. *Science* 195: 260–262.

Schlesinger, W. H., L. A. Bruijnzeel, M. B. Bush, E. M Klein, K. A. Mace, J. A. Raikes, and R. J. Whittaker. 1998. The biogeochemistry of phosphorus after the first century of soil development on Rakata Island, Krakatau, Indonesia. *Biogeochemistry* 40: 37–55.

Schuur, E.A.G. 2001. The effect of water on decomposition dynamics in mesic to wet Hawaiian montane forests. *Ecosystems* 4: 259–273.

———. 2003. Net primary productivity and global climate revisited: the sensitivity of tropical forest growth to precipitation. *Ecology* 84: 1165–1170.

Schuur, E.A.G., and P. A. Matson. 2001. Net primary productivity and nutrient cycling across a mesic to wet precipitation gradient in Hawaiian montane forests. *Oecologia* 128: 431–442.

Schuur, E.A.G., O. A. Chadwick, and P. A. Matson. 2001. Carbon cycling and soil carbon storage in mesic to wet Hawaiian montane forests. *Ecology* 82: 3182–3196.

Schwertmann, U., and R. M. Taylor. 1989. Iron oxides. In *Minerals in Soil Environments (Second Edition)*, eds. J. B. Dixon, and S. B. Weed, 379–438. Madison, Wisconsin: Soil Science Society of America.

Scowcroft, P. G. 1997. Mass and nutrient dynamics of decaying litter from *Passiflora mollissima* and selected native species in a Hawaiian montane rainforest. *Journal of Tropical Ecology* 13: 539–558.

Scowcroft, P. G., D. R. Turner, and P. M. Vitousek. 2000. Decomposition of *Metrosideros polymorpha* leaf litter along elevational gradients in Hawaii. *Global Change Biology* 6: 73–85.

Sherman, G. D., and H. Ikawa. 1968. Soil sequences in the Hawaiian Islands. *Pacific Science* 22: 458–464.

Shoji, S., M. Nanzyo, and R. A. Dahlgren. 1993. *Volcanic Ash Soils*. Amsterdam: Elsevier.

Sigman, D. M., and E. A. Boyle. 2000. Glacial/inter-glacial variations in atmospheric carbon dioxide. *Nature* 407: 859–869.

Silver, W. L., and R. K. Miya. 2001. Global patterns in root decomposition: comparisons of climate and litter quality effects. *Oecologia* 129: 407–419.

Silver, W. L., F. N. Scatena, A. H. Johnson, T. G. Siccama, and M. J. Sanchez. 1994. Nutrient availability in a montane wet tropical forest: Spatial patterns and methodological considerations. *Plant and Soil* 164: 129–145.

Silvester, W. B. 1989. Molybdenum limitation of asymbiotic nitrogen fixation in forests of Pacific Northwest America. *Soil Biology and Biochemistry* 21: 283–289.

Simonson, R. W. 1995. Airborne dust and its significance to soils. *Geoderma* 65: 1–43.

Singer, F. J., W. T. Swank, and E.E.C. Clebsch. 1984. Effects of wild pig rooting in a deciduous forest. *Journal of Wildlife Management* 48: 464–473.

Sinsabaugh, R. L. 1994. Enzymic analysis of microbial pattern and process. *Biology and Fertility of Soils* 17: 69–74.

Skottsberg, C. 1941. Plant succession on recent lava flows in the island of Hawaii. *Gotebergs Kungl. Vetenscapsoch Vitterhets-samhalles Handlingar.* Sjatte Folden, Series B, Band 1, No. 8.

Smeck, N. E. 1985. Phosphorus dynamics in soils and landscapes. *Geoderma* 36: 185-199.

Smil, V. 1990. Nitrogen and phosphorus. In *The Earth as Transformed by Human Action*, eds. B. L. Turner II, W. C. Clark, R. W. Kates, J. F. Richards, J. T. Mathews, and W. B. Meyer, 423–436. Cambridge: Cambridge University Press.

———. 2000. Phosphorus in the environment: Natural flows and human interferences. *Annual Review of Energy in the Environment* 25: 53–88.

Smith, V. H. 1992. Effects of nitrogen:phosphorus supply ratios in nitrogen fixation in agricultural and pastoral systems. *Biogeochemistry* 18: 19–35.

Sollins, P., G. P. Robertson, and G. Uehara. 1988. Nutrient mobility in variable- and permanent-charge soils. *Biogeochemistry* 6: 181–199.

Soltis, D. E., P. S. Soltis, D. R. Morgan, S. M. Swenson, B. C. Mullin, J. M. Dowd, and P. G. Martin. 1995. Chloroplast gene sequence data suggest a single origin of the predisposition for symbiotic nitrogen fixation in angiosperms. *Proceedings of the National Academy of Science* 92: 2647–2651.

Sprent, J. I. 1999. Nitrogen fixation and growth of non-crop legume species in diverse environments. *Perspectives in Plant Ecology, Evolution and Systematics* 2: 149–162.

Sprent, J. I., and P. Sprent. 1990. *Nitrogen Fixing Organisms: Pure and Applied Aspects*. London: Chapman & Hall.

Stark, J. M., and S. C. Hart. 1997. High rates of nitrification and nitrate turnover in undisturbed coniferous forests. *Nature* 385: 61–64.

Steadman, D. W. 1995. Prehistoric extinctions of Pacific island birds: biodiversity meets zooarcheology. *Science* 267: 1123–1131.

Stemmermann, L. 1983. Ecological studies of Hawaiian *Metrosideros* in a successional context. *Pacific Science* 37: 361–373.

Sterner, R., and J. J. Elser. 2002. *Ecological Stoichiometry: The Biology of Elements from Molecules to the Biosphere*. Princeton: Princeton University Press.

Stevens, P. R., and T. W. Walker. 1970. The chronosequence concept and soil formation. *Quarterly Review of Biology* 45: 333–350.

Stewart, B. W., R. C. Capo, and O. A. Chadwick. 1998. Quantitative strontium isotope models for weathering, pedogenesis and biogeochemical cycling. *Geoderma* 82: 173–195.

———. 2001. Effects of rainfall on weathering rate, base cation provenance, and Sr isotope composition of Hawaiian soils. *Geochimica et Cosmochimica Acta* 65: 1087–1099.

Stoorvogel, J. J., N. van Breemen, and B. H. Janssen. 1997. The nutrient input by harmattan dust to a forest ecosystem Cote d'Ivoire, Africa. *Biogeochemistry* 37: 145–157.

Sundareshwar, P. V., J. T. Morris, E. K. Koepfler, and B. Fornwalt. 2003. Phosphorus limitation of coastal ecosystem processes. *Science* 299: 563–565.

Swank, W. T., and D. A. Crossley (eds.). 1988. *Forest Hydrology and Ecology at Coweeta*. New York: Springer-Verlag.

Swap, R., M. Garstang, S. Greco, R. Talbot, and P. Kallbert. 1992. Saharan dust in the Amazon Basin. *Tellus* 44B: 133–149.

Swift, M. J., O. W. Heal, and J. M. Anderson. 1979. *Decomposition in Terrestrial Environments*. Oxford: Blackwell Scientific.

Tamm, C. O. 1991. Nitrogen in terrestrial ecosystems. *Ecological Studies* 81: 1–115.

Tanner, E.V.J., V. Kapos, and W. Franco. 1992. Nitrogen phosphorus fertilization effects on Venezuelan montane forest trunk growth and litterfall. *Ecology* 73: 78–86.

Tanner, E.V.J., P. M. Vitousek, and E. Cuevas. 1998. Experimental investigation of nutrient limitation of forest growth on wet tropical mountains. *Ecology* 79: 10–22.

Tateno, M., and F. S. Chapin. 1997. The logic of carbon and nitrogen interactions in terrestrial ecosystems. *American Naturalist* 149: 723–744.

Thompson, M. V., and P. M. Vitousek. 1997. Asymbiotic nitrogen fixation and decomposition during long-term soil development in Hawaiian montane rain forest. *Biotropica* 29: 134–144.

Tiedje, J. M., S. Asuming-Brempong, K. Nusslein, T. L. Marsh, and S. J. Flynn. 1999. Opening the black box of soil microbial diversity. *Applied Soil Ecology* 13: 109–122.

Tiessen, H., and J. O. Moir. 1993. Characterization of available P by sequential extraction. In *Soil Sampling and Methods of Analysis,* ed. M. R. Carter, 75–86. Boca Raton, FL: Louis.

Tilman, D. 1988. *Plant Strategies and the Dynamics and Function of Plant Communities.* Princeton: Princeton University Press.

Tilman, D., D. Wedin, and J. Knops. 1996. Productivity and sustainability influenced by biodiversity in grassland ecosystems. *Nature* 379: 718–720.

Tilman, D., P. Reich, J. Knops, D. Wedin, T. Mielke, and C. Lehmann. 2001. Diversity and productivity in a long-term grassland experiment. *Science* 294: 843–845.

Torn, M. S., S. E. Trumbore, O. A. Chadwick, P. M. Vitousek, and D. M. Hendricks. 1997. Mineral control of soil carbon storage and turnover. *Nature* 389: 170–173.

Torn, M. S., P. M. Vitousek, and S. E. Trumbore. The influence of nutrient availability on soil organic matter turnover estimated by incubation and radiocarbon modeling. Submitted.

Townsend, A. R., P. M. Vitousek, and S. E. Trumbore. 1995. Temperature controls over long-term soil organic matter dynamics. *Ecology* 76: 721–733.

Townsend, A. R., P. M. Vitousek, D. J. Des Marais, and A. Tharpe. 1997. Effects of temperature and soil carbon pool structure on CO_2 and $^{13}CO_2$ fluxes from five Hawaiian soils. *Biogeochemistry* 38: 1–17.

Treseder, K. K., and P. M. Vitousek. 2001a. Effects of soil nutrient availability on investment in acquisition of N and P in Hawaiian rain forests. *Ecology* 82: 946–954.

———. 2001b. Potential ecosystem-level effects of genetic variation among populations of *Metrosideros polymorpha* from a soil fertility gradient in Hawaii. *Oecologia* 126: 266–275.

Trumbore, S. E. 2000. Age of soil organic matter and soil respiration: radiocarbon constraints of belowground C dynamics. *Ecological Applications* 10: 399–411.

Trumbore, S. E., O. A. Chadwick, and R. G. Amundson. 1996. Rapid exchange between soil carbon and atmospheric CO_2 driven by temperature change. *Science* 272: 393–396.

Tyrrell, T. 1999. The relative influences of nitrogen and phosphorus on oceanic primary production. *Nature* 400: 525–531.

Uebersax, A. 1996. The content and stable isotope systematics of carbon and nitrogen in soil organic matter from elevational transects in Hawaii (USA) and Mt. Kilimanjaro (Tanzania). M.S. thesis, University of California, Berkeley.

Uehara, G., and G. Gillman 1981. *The Mineralogy, Chemistry, and Physics of Tropical Soils with Variable Charge Clays.* Boulder, CO: Westview Press.

Van Cleve, K., F. S. Chapin III, C. T. Dyrness, and L. A. Viereck. 1991. Element cycling in taiga forest: State-factor control. *BioScience* 41: 78–88.

Van den Driessche, R. 1974. Prediction of mineral nutrient status of trees by foliar analysis. *Botanical Review* 40: 347–394.

Van Soest, P. J., and R. H. Wine. 1968. Determination of lignin and cellulose in acid-detergent fiber with permanganate. *Journal of the American Association of Analytical Chemists* 51: 780–785.

Van Vuuren, M.M.I., and L. J. van der Eerden. 1992. Effects of three rates of atmospheric nitrogen deposition enriched with ^{15}N on litter decomposition in a heathland. *Soil Biology and Biochemistry* 24: 527–532.

Vitousek, P. M. 1982. Nutrient cycling and nutrient use efficiency. *American Naturalist* 119: 553–572.

———. 1986. Biological invasions and ecosystem properties: can species make a difference? In *Biological Invasions of North America and Hawaii*, eds. H. A. Mooney, and J. Drake, 153–176. New York: Springer-Verlag.

———. 1994. Potential nitrogen fixation during primary succession in Hawaii Volcanoes National Park. *Biotropica* 26: 234–240.

———. 1995. The Hawaiian Islands as a model system for ecosystem studies. *Pacific Science* 49: 2–16.

———. 1998. Foliar and litter nutrients, nutrient resorption, and decomposition in Hawaiian *Metrosideros polymorpha*. *Ecosystems* 1: 401–407.

———. 1999. Nutrient limitation to nitrogen fixation in young volcanic sites. *Ecosystems* 2: 505–510.

———. 2002. Oceanic islands as model systems for ecological studies. *Journal of Biogeography* 29: 1–10.

———. 2003. Stoichiometry and flexibility in the Hawaiian model system. In *Interactions of the Major Biogeochemical Cycles: Global Change and Human Impacts,* eds. J. M. Melillo, C. B. Field, and B. Moldan, 133–177. Washington, DC: Island Press.

Vitousek, P. M., and H. Farrington. 1997. Nutrient limitation and soil development: experimental test of a biogeochemical theory. *Biogeochemistry* 37: 63–75.

Vitousek, P. M., and C. B. Field. 1999. Ecosystem constraints to symbiotic nitrogen fixers: a simple model and its implications. *Biogeochemistry* 46: 179–202.

———. 2001. Input-output balances and nitrogen limitation in terrestrial ecosystems. In *Global Biogeochemical Cycles in the Climate System,* eds. E. D. Schulze, S. P. Harrison, M. Heimann, E. A. Holland, J. Lloyd, I. C. Prentice, and D. Schimel, 217–235. San Diego: Academic Press.

Vitousek, P. M., and S. E. Hobbie. 2000. The control of heterotrophic nitrogen fixation in decomposing litter. *Ecology* 81: 2366–2376.

Vitousek, P. M., and R. W. Howarth. 1991. Nitrogen limitation on land and in the sea: how can it occur? *Biogeochemistry* 13: 87–115.

Vitousek, P. M., and P. A. Matson. 1985. Disturbance, nitrogen availability, and nitrogen losses in an intensively managed loblolly pine plantation. *Ecology* 66: 1360–1376.

Vitousek, P. M. and J. M. Melillo. 1979. Nitrate losses from disturbed ecosystems: patterns and mechanisms. *Forest Science* 25: 605–619.

Vitousek, P. M., and W. A. Reiners. 1975. Ecosystem succession and nutrient retention: a hypothesis. *BioScience* 25: 376–381.

Vitousek, P. M., and R. L. Sanford, Jr. 1986. Nutrient cycling in moist tropical forest. *Annual Review of Ecology and Systematics* 17: 137–167.

Vitousek, P. M., and L. R. Walker. 1989. Biological invasion by *Myrica faya* in Hawai'i: plant demography, nitrogen fixation, and ecosystem effects. *Ecological Monographs* 59: 247–265.

Vitousek, P. M., J. R. Gosz, C. C. Grier, J. M. Melillo, and W. A. Reiners. 1982. A comparative analysis of potential nitrification and nitrate mobility in forest ecosystems. *Ecological Monographs* 52: 155–177.

Vitousek, P. M., L. R. Walker, L. D. Whiteaker, D. Mueller-Dombois, and P. A. Matson. 1987. Biological invasion by *Myrica faya* alters ecosystem development in Hawaii. *Science* 238: 802–804.

Vitousek, P. M., P. A. Matson, C. Volkmann, J. M. Maass, and G. Garcia. 1989. Nitrous oxide flux from seasonally dry tropical forests: a survey. *Global Biogeochemical Cycles* 3: 375–382.

Vitousek, P. M., C. B. Field, and P. A. Matson. 1990. Variation in foliar $\delta^{13}C$ in Hawaiian *Metrosideros polymorpha:* a case of internal resistance? *Oecologia* 84: 362–370.

Vitousek, P. M., G. Aplet, D. R. Turner, and J. J. Lockwood. 1992. The Mauna Loa environmental matrix: foliar and soil nutrients. *Oecologia* 89: 372–382.

Vitousek, P. M., L. R. Walker, L. D. Whiteaker, and P. A. Matson. 1993. Nutrient limitation to plant growth during primary succession in Hawaii Volcanoes National Park. *Biogeochemistry* 23: 197–215.

Vitousek, P. M., D. R. Turner, W. J. Parton, and R. L. Sanford. 1994. Litter decomposition on the Mauna Loa environmental matrix, Hawaii: Patterns, mechanisms, and models. *Ecology* 75: 418–429.

Vitousek, P. M., G. H. Aplet, J. R. Raich, and J. P. Lockwood. 1995a. Mauna Loa as a model system for ecological research. *Geophysical Monographs* 92: 117–126.

Vitousek, P. M., D. R. Turner, and K. Kitayama. 1995b. Foliar nutrients during long-term soil development in Hawaiian montane rain forest. *Ecology* 76: 712–720.

Vitousek, P. M., J. D. Aber, R. W. Howarth, G. E. Likens, P. A. Matson, D. W. Schindler, W. H. Schlesinger, and D. Tilman. 1997a. Human alteration of the global nitrogen cycle: sources and consequences. *Ecological Applications* 7: 737–750.

Vitousek, P. M., O. A. Chadwick, T. Crews, J. Fownes, D. Hendricks, and D. Herbert. 1997b. Soil and ecosystem development across the Hawaiian Islands. *GSA Today* 7(9): 1–8.

Vitousek, P. M., L. O. Hedin, P. A. Matson, J. H. Fownes, and J. Neff. 1998. Within-system element cycles, input-output budgets, and nutrient limitation. In *Successes, Limitations, and Frontiers in Ecosystem Science,* eds. M. Pace, and P. Groffman, 432–451. New York: Springer-Verlag.

Vitousek, P. M., M. J. Kennedy, L. A. Derry, and O. A. Chadwick. 1999. Weathering versus atmospheric sources of strontium in ecosystems on young volcanic soils. *Oecologia* 121: 255–259.

Vitousek, P. M., K. Cassman, C. Cleveland, T. Crews, C. B. Field, N. B. Grimm, R. W. Howarth, R. Marino, L. Martinelli, E. B. Rastetter, and J. I. Sprent. 2002. Towards an ecological understanding of biological nitrogen fixation. *Biogeochemistry* 57: 1–45.

Vitousek, P. M., O. A. Chadwick, P. A. Matson, S. Allison, L. A. Derry, L. Kettley, A. Luers, E. Mecking, V. Monastra, and S. Porder. Erosion and the rejuvenation of weathering-derived nutrient supply in an old tropical landscape. *Ecosystems,* in press.

Wada, K. 1989. Allophane and imogolite. In *Minerals in Soil Environments (Second Edition),* eds. J. B. Dixon, and S. B. Weed, 1051–1087. Madison, WI: Soil Science Society of America.

Wagner, W. L., D. R. Herbst, and S. H. Sohmer. 1990. *Manual of Flowering Plants of Hawaii.* Honolulu: B. P. Bishop Museum.

Walker, J., C. H. Thompson, and W. Jehne. 1983. Soil weathering stage, vegetation succession, and canopy dieback. *Pacific Science* 37: 471–481.

Walker, L. R., and G. H. Aplet. 1994. Growth and fertilization responses of Hawaiian tree ferns. *Biotropica* 26: 378–383.

Walker, L. R., and P. M. Vitousek. 1991. Interactions of an alien and a native tree during primary succession in Hawaii Volcanoes National Park. *Ecology* 72: 1449–1455.

Walker, L. R., D. J. Zarin, N. Fetcher, R. W. Myster, and A. H. Johnson. 1996. Ecosystem development and plant succession on landslides in the Caribbean. *Biotropica* 28: 566–576.

Walker, T. W., and J. K. Syers. 1976. The fate of phosphorus during pedogenesis. *Geoderma* 15: 1–19.

Waksman, S. A., and F. G. Tenney. 1928. Composition of natural organic materials and their decomposition in the soil. III. The influence of nature of plant upon the rapidity of its decomposition. *Soil Science* 26: 155–171.

Wardle, D. A. 1999. Is "sampling effect" a problem for experiments investigation biodiversity-ecosystem function relationships? *Oikos* 87: 403–407.

———. 2002. *Communities and Ecosystems: Linking the Aboveground and Belowground Components.* Princeton: Princeton University Press.

Weathers, K. C., and G. E. Likens. 1997. Clouds in southern Chile: an important source of nitrogen to nitrogen-limited ecosystems? *Environmental Science and Technology* 31: 210–213.

Weathers, K. C., G. M. Lovett, G. E. Likens, and N.F.M. Caraco. 2000. Cloudwater inputs of nitrogen to forest ecosystems in southern Chile: Forms, fluxes, and sources. *Ecosystems* 3: 590–595.

Wedin, D. A., and D. Tilman. 1990. Species effects on nitrogen cycling: A test with perennial grasses. *Oecologia* 84: 433–441.

White, T.C.R. 1993. *The Inadequate Environment: Nitrogen and the Abundance of Animals.* Berlin: Springer-Verlag.

Wolfe, E. W., and J. Morris. 1996. Geologic map of the island of Hawaii. Reston, Virginia: U.S. Geological Survey Map I-2524-A, Miscellaneous Investigations Series.

Wood, T., F. H. Bormann, and G. K. Voigt. 1984. Phosphorus cycling in a northern hardwood forest: Biological and geochemical control. *Science* 223: 391–393.

Wright, S. D., C. G. Yong, J. W. Dawson, D. J. Whittaker, and R. C. Gardner. 2000. Riding the ice age El Nino? Pacific biogeography and evolution of *Metrosideros* subg. *Metrosideros* (Myrtaceae) inferred from nuclear ribosomal DNA. *Proceedings of the National Academy of Sciences* 97: 4118–4123.

Wright, S. D., C. G. Yong, S. R. Wichman, J. W. Dawson, and R. C. Gardner. 2001. Stepping stones to Hawai'i: A transequatorial dispersal pathway for Metrosideros (Myrtaceae) inferred from mDNA (ITS + ETS). *Journal of Biogeography* 28: 769–774.

Wright, T. 1971. Chemistry of Kilauea and Mauna Loa lavas in space and time. U.S. Geological Survey Professional Paper 735: 1–49.

Wright, T., and R. T. Helz. 1987. Recent advances in Hawaiian petrology and geochemistry. In *Volcanism in Hawaii,* eds. R. W. Decker, T. L. Wright, and P. M. Stauffer, 625–640. USGS Professional Paper 1350, U.S. Geological Survey, Washington, DC.

Yamagata, Y., H. Watanabe, M. Saitoh, and T. Namba. 1991. Volcanic production of polyphosphates and its relevance to prebiotic evolution. *Nature* 352: 516–519.

Yost, R. S., G. Uehara, and R. L. Fox. 1982. Geostatistical analyses of soil chemical properties of large land areas. *Soil Science Society of America Journal* 46: 1028–1037.

Zarin, D. J., and A. H. Johnson. 1995. Nutrient accumulation during primary succession in a montane tropical forest, Puerto Rico. *Soil Science Society of America Journal* 59: 1444–1452.

Zhang, L., S. Gong, J. Padro, and L. Barrie. 2001. A size-segregated particle dry deposition scheme for an atmospheric aerosol module. *Atmospheric Environment* 35: 549–560.

Ziegler, A. C. 2002. *Hawaiian Natural History, Ecology, and Evolution.* Honolulu: University of Hawai'i Press.

INDEX